Lecture Notes in Computer Science   12435

More information about this series at http://www.springer.com/series/7409

Zhisheng Huang · Siuly Siuly ·
Hua Wang · Rui Zhou ·
Yanchun Zhang (Eds.)

Health
Information Science

9th International Conference, HIS 2020
Amsterdam, The Netherlands, October 20–23, 2020
Proceedings

 Springer

Editors
Zhisheng Huang
Vrije University of Amsterdam
Amsterdam, The Netherlands

Hua Wang 🆔
Victoria University
Footscray, VIC, Australia

Yanchun Zhang 🆔
Victoria University
Footscray, VIC, Australia

Siuly Siuly 🆔
Victoria University
Footscray, VIC, Australia

Rui Zhou 🆔
Swinburne University of Technology
Hawthorn, VIC, Australia

ISSN 0302-9743   ISSN 1611-3349   (electronic)
Lecture Notes in Computer Science
ISBN 978-3-030-61950-3   ISBN 978-3-030-61951-0   (eBook)
https://doi.org/10.1007/978-3-030-61951-0

LNCS Sublibrary: SL3 – Information Systems and Applications, incl. Internet/Web, and HCI

This Springer imprint is published by the registered company Springer Nature Switzerland AG
The registered company address is: Gewerbestrasse 11, 6330 Cham, Switzerland

Preface

The International Conference Series on Health Information Science (HIS) provides a forum for disseminating and exchanging multidisciplinary research results in computer science/information technology and health science and services. It covers all aspects of health information sciences and systems that support health information management and health service delivery.

The 9th International Conference on Health Information Science (HIS 2020) was held in Amsterdam and Leiden, The Netherlands, during October 20–23, 2020. Founded in April 2012 as the International Conference on Health Information Science and Their Applications, the conference continues to grow to include an ever broader scope of activities. The main goal of these events is to provide international scientific forums for researchers to exchange new ideas in a number of fields that interact in-depth through discussions with their peers from around the world. The scope of the conference includes: (1) medical/health/biomedicine information resources, such as patient medical records, devices, and equipment, software and tools to capture, store, retrieve, process, analyze, and optimize the use of information in the health domain; (2) data management, data mining, and knowledge discovery, all of which play a key role in decision-making, management of public health, examination of standards, privacy, and security issues; (3) computer visualization and artificial intelligence for computer-aided diagnosis; and (4) development of new architectures and applications for health information systems.

The conference solicited and gathered technical research submissions related to all aspects of the conference scope. All the submitted papers were peer-reviewed by at least three international experts drawn from the Program Committee. After the rigorous peer-review process, a total of 11 full papers and 6 short papers among 62 submissions were selected on the basis of originality, significance, and clarity and were accepted for publication in the proceedings. The authors were from Australia, China, Finland, India, The Netherlands, and the UK. 8 papers were invited to submit to a special issue of the *Health Information Science and System* journal and will be published by Springer.

The high quality of the program – guaranteed by the presence of an unparalleled number of internationally recognized top experts – is reflected in the content of the proceedings. The conference was therefore a unique event, where attendees were able to appreciate the latest results in their field of expertise, and to acquire additional knowledge in other fields. The program was structured to favor interactions among attendees coming from many different areas, scientifically and geographically, from academia and from industry.

Our thanks go to the host organization, Vrije Universiteit Amsterdam, The Netherlands. Finally, we acknowledge all those who contributed to the success of HIS 2020 but whose names are not listed here.

October 2020

Zhisheng Huang
Siuly Siuly
Hua Wang
Rui Zhou
Yanchun Zhang

Organization

General Co-chairs

Uwe Aickelin	The University of Melbourne, Australia
Yanchun Zhang	Victoria University, Australia
Frank van Harmelen	Vrije Universiteit Amsterdam, The Netherlands

Program Co-chairs

Zhisheng Huang	Vrije Universiteit Amsterdam, The Netherlands
Siuly Siuly	Victoria University, Australia
Hua Wang	Victoria University, Australia

Sponsor Chair

Ran Dang	Atlantis Press, France

Publication Chair

Rui Zhou	Swinburne University of Technology, Australia

Local Arrangement Co-chairs

Ting Liu	Vrije Universiteit Amsterdam, The Netherlands
Xu Wang	Vrije Universiteit Amsterdam, The Netherlands

Workshop Co-chairs

Rui Zhou	Swinburne University of Technology, Australia
Haiyuan Wang	Ztone Beijing, China

Website Co-chairs

Haiyuan Wang	Ztone Beijing, China
Di Wang	Ztone Beijing, China

Finance Chair

Qing Hu	Ztone International BV, The Netherlands

Program Committee

Surpriya Angra	Victoria University, Australia
Jinli Cao	La Trobe University, Australia
Soon Ae Chun	The City University of New York, USA
Licong Cui	The University of Texas Health Science Center at Houston, USA
Yanhui Guo	University of Illinois Springfield, USA
Marcos Gutierrez Alves	Zhejiang University, Ningbo Institute of Technology, China
Xinpeng Jiang	Central China Normal University, China
Xia Jing	Clemson University, USA
Enamul Kabir	University of Southern Queensland, Australia
Rui Li	Xi'an Jiaotong University, China
Shaofu Lin	Beijing University of Technology, China
Gang Luo	University of Washington, USA
Yue Ma	LRI-CNRS, Université Paris Sud, France
William Song	Dalarna University, Sweden
Weiqing Sun	University of Toledo, USA
Xiaohui Tao	University of Southern Queensland, Australia
Annette Ten Teije	Vrije Universiteit Amsterdam, The Netherlands
Ye Wang	National University of Defense Technology, China
Yimin Wen	Guilin University of Electronic Technology, China
Juanying Xie	Shaanxi Normal University, China
Dan Xie	Hubei University of Chinese Medicine, China
Xiaofei Yang	Xi'an Jiaotong University, China
Bingxiang Yang	Wuhan University, China
Xiaolong Zheng	Chinese Academy of Sciences, China
Youwen Zhu	Nanjing University of Aeronautics and Astronautics, China

Additional Reviewers

Tian, Hu	Jahan, Farha
Fu, Guanghui	Pujari, Medha
Fu, Chengcheng	Tong, Jizhou
Abeysinghe, Rashmie	Zhang, Xingwei
Zheng, Fengbo	Wang, Yujie
Abidi, Aman	Malik, Shahida
Sun, Xia	

Contents

Mental Health

Making Sense of Violence Risk Predictions Using Clinical Notes

Pablo Mosteiro[1(✉)], Emil Rijcken[1,2], Kalliopi Zervanou[2], Uzay Kaymak[2], Floortje Scheepers[3], and Marco Spruit[1]

[1] Utrecht University, Utrecht, The Netherlands
{p.mosteiro,m.r.spruit}@uu.nl
[2] Eindhoven University of Technology, Eindhoven, The Netherlands
{e.f.g.rijcken,k.zervanou,u.kaymak}@tue.nl
[3] University Medical Center Utrecht, Utrecht, The Netherlands
f.e.scheepers-2@umcutrecht.nl

Abstract. Violence risk assessment in psychiatric institutions enables interventions to avoid violence incidents. Clinical notes written by practitioners and available in electronic health records (EHR) are valuable resources that are seldom used to their full potential. Previous studies have attempted to assess violence risk in psychiatric patients using such notes, with acceptable performance. However, they do not explain *why* classification works and how it can be improved. We explore two methods to better understand the quality of a classifier in the context of clinical note analysis: random forests using topic models, and choice of evaluation metric. These methods allow us to understand both our data and our methodology more profoundly, setting up the groundwork for improved models that build upon this understanding. This is particularly important when it comes to the generalizability of evaluated classifiers to new data, a trustworthiness problem that is of great interest due to the increased availability of new data in electronic format.

Keywords: Natural Language Processing · Topic modeling · Electronic Health Records · Interpretability · Document classification · LDA · Random forests

1 Introduction

Two thirds of mental health professionals working in Dutch clinical psychiatry institutions report having been a victim of at least one physical violence incident in their careers [14]. These incidents can have a strong psychological effect on nurses [11], as well as economical consequences [20]. Multiple Violence Risk Assessment (VRA) approaches have been proposed to predict and avoid violence incidents, with some adoption in practice [27]. One common approach is the Brøset Violence Checklist (BVC) [2], a questionnaire used by nurses and psychiatrists to evaluate the likelihood for a patient to become involved in a

© Springer Nature Switzerland AG 2020
Z. Huang et al. (Eds.): HIS 2020, LNCS 12435, pp. 3–14, 2020.
https://doi.org/10.1007/978-3-030-61951-0_1

violence incident. This is a time-consuming and highly subjective process, and automation would be beneficial to the field.

Machine learning methods have been successfully applied to psychiatric Electronic Health Records (EHR) to predict readmission [25]. Most current applications of text processing in psychiatric EHR's are for the English language [12]. Building up on promising first attempts to systematically analyse EHR's in Dutch [17–19], the COVIDA project (COmputing VIsits DAta) aims to create a publicly available self-service facility for Natural Language Processing (NLP) of Dutch medical texts.

In order to build a self-service tool, it is essential to dig deep into the machine learning methods employed and build confidence and trust in practitioners. In this work, we investigate VRA using clinical notes in Dutch and attempt to provide better understanding and interpretable results. For this purpose, we re-implement an SVM document classification approach suggested in [17,19] on a 35% bigger data set; we expand on the text features used by combining text with existing structured data, such as the patient's age; and we experiment with alternative document representation techniques, such as LDA and word embeddings, using random forest classification in order to also gain better insights in potentially significant features. We find promising results, though much work remains to be done to achieve acceptable performance for the clinical practice.

2   Related Work

The analysis of free text in EHR's and the combination of these to structured data using machine learning approaches is gaining an increasing interest as anonymised EHR's become available for research. However, the analysis of clinical free-text data presents numerous challenges due to (i) *highly imbalanced data* with respect to the class of interest [23]; (ii) *lack of publicly available data sets*, limiting research on private institutional data [30]; and (iii) relatively *small data sizes* compared to the amounts of data currently used in text processing research.

In the psychiatric domain, structured data such as symptom codes and medication history have been used for the prediction of admissions [9,15]. Studies using structured information in EHR's to predict suicide risk [6] indicate that information from the unstructured data in clinical texts may provide better insights on risk factors and result in better predictions. Free text in combination with structured EHR variables has been used in suicide [7] and depression diagnosis [10] among healthy and unhealthy individuals. In such approaches, structured variables such as medication history, questionnaires, and demographics are expected to provide enough discriminatory power for the required analysis. Research approaches focusing on text from EHR's in mental healthcare are to our knowledge very few; Poulin *et al.* [21] attempted to predict suicide risk among veterans and more recently Menger *et al.* [19] used Dutch clinical text to predict violent incidents from patients in treatment facilities.

The most popular machine learning methods used for processing free text in psychiatric EHR data are support vector machines (SVM), logistic regression, naive Bayes, and decision trees [1]. Decision-tree classification is one of the most easily interpretable approaches, because it allows for inspection of the specific feature combination used for the classification. This line of classification approaches has also achieved significant improvements in classification accuracy by growing an ensemble of decision trees (a *random forest*) trained on subsets of the data set and letting them vote for the most popular class [4].

3 Data Set

The data used in this study consists of clinical notes written in Dutch by nurses and physicians about patients in the psychiatry ward of the University Medical Center (UMC) Utrecht between 2012-08-01 and 2020-03-01. The 834834 notes available are de-identified for patient privacy using DEDUCE [18].

Each patient can be admitted to the psychiatry ward multiple times. In addition, an admitted patient can spent time in various sub-departments of psychiatry. The time the patient spends in each of the sub-departments is called an *admission period*. In the present study, our data points are admission periods. For each admission period, all notes collected between 28 days before and 1 day after the start of the admission period are concatenated and considered as a single *period note*. If a patient is involved in a violence incident between 1 and 28 days after the start of the admission period, the outcome is recorded as *violent* (hereafter also referred to as *positive*). Otherwise, it is recorded as *non-violent* (*i.e.*, *negative*). Admission periods having period notes with fewer than or equal to 100 words are discarded as was done in previous work [19, 25].

In addition to notes, we employ structured variables collected in various formats by the hospital. These include variables related to:

– Admission periods (*e.g.*, start date and time)
– Notes (*e.g.*, date and time of first & last notes in period)
– Patient (*e.g.*, gender, age at the start of the admission period)
– Medications (*e.g.*, numbers prescribed and administered)
– Diagnoses (*e.g.*, presence or absence)

These are included to establish whether some of these variables can be correlated with violence incidents.

The resulting data set consists of 4280 admission periods, corresponding to 2892 unique patients. The data set is highly imbalanced, as a mere 425 admission periods have a violent outcome. In further sections, we will discuss how the imbalanced nature of the data set affects the analysis.

4 Methodology

In this work, we address the problem of violence risk prediction as a document classification task, where EHR document features are combined with additional structured data, as explained in Sect. 3. For text normalization purposes, we perform a series of pre-processing steps outlined in Sect. 4.1. Then, for document representation purposes, we experiment with two alternative approaches—paragraph embeddings and LDA topic vectors—discussed in Sect. 4.2. For the classification task, we experiment with SVM [8] and random forest classification [4] (Sect. 4.3). Finally, we discuss our choice of evaluation metrics in Sect. 4.4.

4.1   Text Normalization

All notes are pre-processed by applying the following normalization steps:

- Converting all period notes to lowercase
- Removing special characters (*e.g.*, ë → e)
- Removing non-alphanumeric characters
- Tokenizing the texts using the NLTK Dutch word tokenizer [3]
- Removing stopwords using the default NLTK Dutch stopwords list
- Stemming using the NLTK Dutch snowball stemmer
- Removing periods

4.2   Text Representations

The language used in clinical text is domain-specific, and the notes are rich in technical terms and spelling errors. Pre-trained paragraph embedding models do not necessarily yield useful representations. For this reason, we use the entire available set of 834834 de-identified clinical notes to train both the paragraph embedding model and the topic model. Only notes with at least 10 words each are used, to remove notes that contain no valuable information.

Paragraph Embeddings. We use Doc2Vec [13] to convert texts to paragraph embeddings. The Doc2Vec training parameters are set to the default values in Gensim 3.8.1 [22], with the exception of four parameters: we increase `epochs` from 5 to 20 to improve the probability of convergence; we increase `min_count`—the minimum number of times a word has to appear in the corpus in order to be considered—from 5 to 20 to avoid including repeated mis-spellings of words [19]; we increase `vector_size` from 100 to 300 to enrich the vectors while keeping the training time acceptable; and we decrease `window`—the size of the context window— from 5 to 2 to mitigate the effects of the lack of structure often present in EHR texts.

Topic Modeling. A previous study using Latent Dirichlet Allocation (LDA) for topic modeling in the psychiatry domain [25] suggests that numeric representations of texts obtained by topic modeling can be used alternatively or in addition to text embeddings in classification problems. We use the LdaMallet [22] implementation of LDA to train a topic model on our large set of 834834 clinical notes. In order to determine the optimal number of topics, we use the coherence model implemented in Gensim to compute the coherence metric [29].

We find that using 25 topics maximizes coherence. We use default values for the LdaMallet training parameters. Using the trained LDA topic model we compute, for each of the 4280 period notes in our data set, a 25-dimensional vector of weights, where each dimension represents a topic and the value represents the degree to which this topic is expressed in the note. These vectors are then used as input to the classifiers described below.

4.3   Classification Methods

Similarly to previous work [17,19], we use Support Vector Machines (SVM) [8]. Moreover, for interpretability purposes, we implement in this work random forest classification [4]. Random forests are ensemble models that are widely used in classification problems [31]. The `scikit-learn` implementation of random forests outputs after training a list of the most relevant features used for classification. This can help us determine whether some of the features are more important than others when it comes to classifying positive and negative samples.

We use two loops of 5-fold cross-validation for estimation of uncertainty and hyper-parameter tuning. In each iteration of the outer loop, the admission periods corresponding to 1/5 of the patients are kept as test data, and the remaining admission periods are used in the inner loop to perform a grid search for hyper-parameter tuning. The best classifier from the inner loop is applied to the test data, and the resulting classification metrics from each iteration of the outer loop are used to calculate a mean and a standard deviation for the metrics.

We employ the SVC support-vector classifier provided by `scikit-learn`, with default parameters except for the following: `class_weight` is set to 'balanced' to account for our imbalanced data set; `probability` is True to enable probability estimates for performance evaluation; the cost parameter C and the kernel coefficient `gamma` are determined by cross-validation. The ranges of values used are C = $\{10^{-1}, 10^0, 10^1\}$ and `gamma` = $\{10^{-5}, 10^{-4}, 10^{-3}, 10^{-2}, 10^{-1}, 10^0\}$. Both of these ranges were motivated in a previous study [19].

For the random forest classifier, we use the `scikit-learn` implementation, with default values for all the parameters except for the following: `n_estimators` is increased to 500 to prevent overfitting; `class_weight` is set to 'balanced' to account for the imbalanced data set; and `min_samples_leaf`, `max_features` and `criterion` are determined by cross-validation. Values for `min_samples_leaf` are greater than the default value of 1, to prevent overfitting. For `max_features`, we consider the default value of 'auto', which sets the maximum number of features per split to the square root of the number of features, and two smaller values, again in order to prevent overfitting. Finally, both split criteria available in `scikit-learn` were considered ('gini' and 'entropy'). These parameters are summarized in Table 1.

4.4 Evaluation Metrics

Binary classifiers predict probabilities for input samples to belong to the positive class. When employing a binary classifier in practice, a threshold is chosen, and all samples with positive probabilities above that threshold are considered positive *predictions*. While testing the performance of a classifier, then, we can compare the actual *conditions* with the predictions, and classify each sample as a *true positive* (TP), *true negative* (TN), *false positive* (FP) (condition negative, predicted positive) or *false negative* (FN) (condition positive, predicted negative).

Table 1. Random forest training parameters. Parameters with multiple values are optimized through cross-validation. Parameters not shown are set to default `scikit-learn` values.

Parameter	Value/s	Method
`min_samples_leaf`	{3, 5, 10}	Cross-validation
`max_features`	{5.2, 8.7, 'auto'}	Cross-validation
`criterion`	{'gini', 'entropy'}	Cross-validation
`n_estimators`	500	Fixed
`class_weight`	'balanced'	Fixed

Choosing an operating threshold in practice requires domain expert knowledge. *E.g.*, if violence incidents have very high costs (human or economic), avoiding false negatives would be a priority; if, on the other hand, interventions are costly and cannot be afforded for most patients, avoiding false positives would be more important. Because the decision of the operating threshold is usually made only when the classifier will be put into practice, we report the performance of classifiers using metrics that are agnostic to the operating threshold.

The performance of classifiers is often reported in terms of the Receiver Operating Characteristic Area Under the Curve (ROC-AUC) [17,19]. The ROC is a plot of the true positive rate (TPR) as a function of the false positive rate (FPR), where TPR and FPR are defined as

$$\text{TPR} = \frac{\text{TP}}{\text{TP} + \text{FN}}; \text{ FPR} = \frac{\text{FP}}{\text{TN} + \text{FP}} \qquad (1)$$

In other words, the curve is constructed by choosing multiple classification thresholds, computing the TP, FP, TN, FN for each threshold, then computing FPR and TPR. As you vary the classification threshold, you allow more or fewer positive predictions, so FPR and TPR both vary in the same direction. In a random classifier, FPR and TPR vary at the same rate, so the baseline ROC is a straight line between (0, 0) and (1, 1), and the baseline ROC-AUC is 0.5. The maximum ROC-AUC is 1, which represents perfect discrimination between TP and FP.

It has been previously noted [26] that ROC-AUC is not a robust performance metric when dealing with imbalanced data sets such as ours. Because the data set is highly imbalanced, the FPR can be misleadingly small, simply because the denominator includes all negative samples, and this artificially increases the ROC-AUC. For this reason, we opt in this work to implement the area under the Precision-Recall curve (PR-AUC) evaluation measure [16]. The Precision-Recall curve is, as its name suggests, a plot of the precision of the classifier as a function of its recall, with precision and recall defined as:

$$\text{Precision} = \frac{\text{TP}}{\text{TP} + \text{FP}}; \ \text{Recall} = \frac{\text{TP}}{\text{TP} + \text{FN}} \tag{2}$$

Note that neither of these quantities are directly dependent on TN, which is a desirable feature because we have an imbalanced data set with a large number of negatives, and we are more interested in the few positives.

To determine the baseline value for PR-AUC, note that, no matter the recall, the best precision that can be achieved by guessing randomly is the real fraction of positive samples, f_P. Thus, the baseline PR-AUC is f_P, which in our case is $425/4280 = 0.10$.

Though we make no decision regarding the classification threshold, we believe that due to the nature of violence incidents it is more important to avoid FN than to avoid FP. Thus, a good metric to quantify the performance of a classifier in practice is F_2, given by:

$$F_\beta = (1 + \beta^2) \cdot \frac{\text{Precision} \cdot \text{Recall}}{\beta^2 \cdot \text{Precision} + \text{Recall}} \tag{3}$$

with $\beta = 2$ [24].

In this work, we report our classifier performance in both PR-AUC and ROC-AUC, for comparison with previous work on similar data sets [17,19,25,28]. We also report $F_2^{\max}$, i.e., the value of F_2 at the classification threshold that maximizes F_2.

5   Experimental Results and Discussion

5.1   Classifier Performance

Table 2 reports the results of the analyses. All configurations gave results consistent with each other, as well as with previous work on a smaller data set [19]. Figure 1 shows the precision-recall curve for one of the folds of the outer uncertainty-estimation loop during the training of the SVM classifier. These metrics show modest performance, and they indicate that further work is needed to extract all the meaningful information contained in the clinical notes.

Table 2. Classification metrics for various training configurations.

LDA	Embeddings	Structured vars	Estimator	PR-AUC	ROC-AUC	$F_2^{\max}$
No	Yes	No	SVM	0.321 ± 0.067	0.792 ± 0.011	0.519
No	Yes	No	RF	0.293 ± 0.054	0.782 ± 0.011	0.514
No	Yes	Yes	RF	0.299 ± 0.056	0.782 ± 0.011	0.515
Yes	No	Yes	RF	0.309 ± 0.070	0.785 ± 0.011	0.503
Yes	Yes	Yes	RF	0.304 ± 0.058	0.792 ± 0.011	0.517

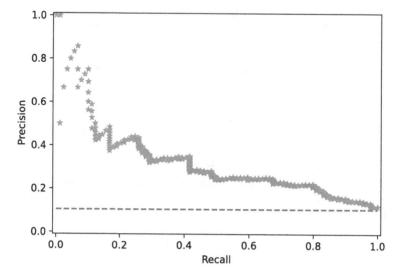

Fig. 1. Precision-Recall curve for one of the folds of the uncertainty-estimation loop during training of the SVM classifier. The PR-AUC is 0.33.

5.2 Feature Importance

When using the random forest estimator, at each step in the outer cross-validation loop we stored the 10 most important features according to the best fit in the inner cross-validation loop. Gathering all the most important features together, we then studied both the 10 most repeated features and the 10 features with the highest total feature importance. These lists were reassuringly similar. The most repeated features were 5 of the text embedding features, plus the age at the beginning of the admission period (age_admission) and the number of words in the period note (num_words). The frequency distributions of these variables are shown, for both positive and negative samples, on Fig. 2. As can be seen in the figures, the average violent patient is younger than the average non-violent patient, and the average period note about a violent patient is longer than the average period note about a non-violent patient.

The fact that only two of the structured variables included in our study resulted in significant discrimination between the positive and negative classes further stresses that novel sophisticated methods are required.

5.3   Inter-classifier Agreement

To better understand why paragraph embeddings and topic models gave similar classification metrics, we studied the inter-classifier agreement using Cohen's kappa [5]. This metric quantifies how much two classifiers agree, taking into consideration the probability that they agree by chance. A value of Cohen's kappa equal to 0 means the agreement between the two classifiers is random, while a value of 1 means the agreement is perfect and non-random. We placed classification thresholds at multiple points between 0 and 1 (same threshold for both classifiers), and computed classification labels for each classifier for each threshold. Using those classification labels, we computed Cohen's kappa. The result is shown in Fig. 3.

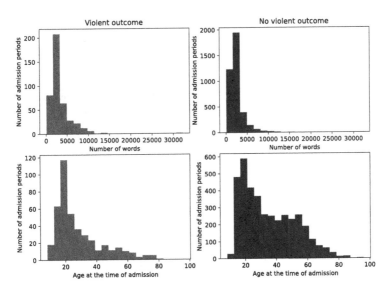

Fig. 2. Histograms of number of words per period note (top) and age (bottom) for violent (left/red) and non-violent (right/blue) patients. (Color figure online)

Next, for each classifier, we used the threshold that maximized the F_2 metric (see Sect. 4.4), and calculated classification labels for each classifier using those thresholds. We then computed Cohen's kappa using those classification labels, and obtained a value of:

$$\kappa = 0.633 \pm 0.012, \tag{4}$$

which is close to the maximum value reported in Fig. 3.

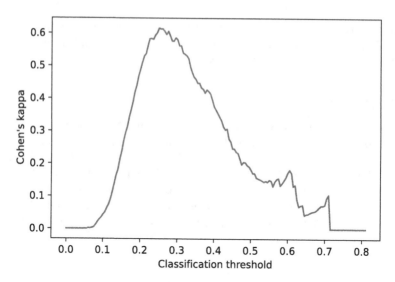

Fig. 3. Cohen's κ score for inter-classifier reliability, comparing the LDA-based classifier and the embeddings-based classifier, as a function of the classification threshold. The same 200 equidistant thresholds between 0 and 1 were used for both classifiers.

According to the standard interpretation of Cohen's kappa, this value implies that the agreement between the two classifiers is better than random, though not close to perfect agreement. This might indicate that the paragraph embeddings and the representations based on LDA are capturing similar information from the notes.

6   Conclusions

We applied machine learning methods to Dutch clinical notes from the psychiatry department of the University Medical Center (UMC) in Utrecht, the Netherlands. We trained a model that predicts, based on the content of the notes, which patients are likely to be involved in a violence incident within their first 4 weeks of admission. The performance of our classifiers is assessed using the area under the precision-recall curve (PR-AUC), and its value is approximately 0.3, well above the baseline value of around 0.1; the ROC-AUC is approximately 0.8. The maximum F_2 score, which puts twice as much importance on recall as on precision, is approximately 0.5. Our results are competitive with a study based on structured variables that obtained ROC-AUC = 0.7801 [28]. These metrics show modest performance, and they indicate that further work is needed to extract all the meaningful information contained in the clinical notes.

The fact that only two of the structured variables included in our study—number of words and patient age—resulted in significant differentiation between the positive and negative classes further stresses that novel sophisticated methods are required. In particular, deep learning is a promising approach.

We have also, for the first time as far as we are aware, applied topic modeling to clinical notes in Dutch language for Violence Risk Assessment. We found that the performance of classifiers on numerical representations produced by topic models is comparable to the performance of similar classifiers on document embeddings. Since topic models are easier to interpret than paragraph embeddings, this is a promising avenue for implementation in practice. However, this is a preliminary study, and more extensive research on larger data sets should be performed to confirm our findings. This is the direction we will follow for future research.

References

1. Abbe, A., Grouin, C., Zweigenbaum, P., Falissard, B.: Text mining applications in psychiatry: a systematic literature review. MPR **25**(2), 86–100 (2016)
2. Almvik, R., Woods, P., Rasmussen, K.: The Brøset violence checklist: sensitivity, specificity, and interrater reliability. J. Interpers. Violence **15**(12), 1284–1296 (2000)
3. Bird, S., Klein, E., Loper, E.: Natural Language Processing with Python: Analyzing Text with the Natural Language Toolkit. https://www.nltk.org/book/
4. Breiman, L.: Random forests. Mach. Learn. **45**, 5–32 (2001)
5. Cohen, J.: A coefficient of agreement for nominal scales. Educ. Psychol. Meas. **20**(1), 37–46 (1960)
6. Conner, K.R., et al.: Mental disorder comorbidity and suicide among 2.96 million men receiving care in the veterans health administration health system. J. Abnorm. Psychol. **122**(1), 256–263 (2012)
7. Cook, B.L., Progovac, A.M., Chen, P., Mullin, B., Hou, S., Baca-Garcia, E.: Novel use of natural language processing (NLP) to predict suicidal ideation and psychiatric symptoms in a text-based mental health intervention in Madrid. Comput. Math. Methods Med. **2016**, 8708434 (2016)
8. Cortes, C., Vapnik, V.: Support-vector networks. Mach. Learn. **20**(3), 273–297 (1995)
9. Friedman, S., Margolis, R., David, O.J., Kesselman, M.: Predicting psychiatric admission from an emergency room. J. Nerv. Ment. Dis. **171**(3), 155–158 (1983)
10. Huang, S.H., LePendu, P., Iyer, S.V., Tai-Seale, M., Carrell, D., Shah, N.H.: Toward personalizing treatment for depression: predicting diagnosis and severity. JAMIA **21**(6), 1069–1075 (2014)
11. Inoue, M., Tsukano, K., Muraoka, M., Kaneko, F., Okamura, H.: Psychological impact of verbal abuse and violence by patients on nurses working in psychiatric departments. Psychiatry Clin. Neurosci. **60**, 29–36 (2006)
12. Kim, Y.K. (ed.): Frontiers in Psychiatry: Artificial Intelligence, Precision Medicine, and Other Paradigm Shifts. Springer, Singapore (2019). https://doi.org/10.1007/978-981-32-9721-0
13. Le, Q., Mikolov, T.: Distributed representations of sentences and documents. In: ICML 2014, Beijing, China, pp. 1188–1196. PMLR (2014)
14. van Leeuwen, M., Harte, J.: Violence against mental health care professionals: prevalence, nature and consequences. J. Forensic Psychiatry Psychol. **28**(5), 581–598 (2017)

15. Lyons, J.S., Stutesman, J., Neme, J., Vessey, J.T., O'Mahoney, M.T., Camper, H.J.: Predicting psychiatric emergency admissions and hospital outcome. Medical care **35**(8), 792–800 (1997)
16. Manning, C., Raghavan, P., Schütze, H.: Introduction to Information Retrieval. Cambridge University Press, Cambridge (2008)
17. Menger, V., Scheepers, F., Spruit, M.: Comparing deep learning and classical machine learning approaches for predicting inpatient violence incidents from clinical text. Appl. Sci. **8**(6), 981 (2018)
18. Menger, V., Scheepers, F., van Wijk, L., Spruit, M.: DEDUCE: a pattern matching method for automatic de-identification of Dutch medical text. Telemat. Inform. **35**(4), 727–736 (2018)
19. Menger, V., Spruit, M., van Est, R., Nap, E., Scheepers, F.: Machine learning approach to inpatient violence risk assessment using routinely collected clinical notes in electronic health records. JAMA Netw. Open **2**(7), e196709 (2019)
20. Nijman, H., Bowers, L., Oud, N., Jansen, G.: Psychiatric nurses' experiences with inpatient aggression. Aggress. Behav. **31**(3), 217–227 (2005)
21. Poulin, C., et al.: Predicting the risk of suicide by analyzing the text of clinical notes. PLoS ONE **9**(3), e91602 (2014)
22. Řehůřek, R., Sojka, P.: Software framework for topic modelling with large corpora. In: Proceedings of LREC 2010 Workshop on New Challenges for NLP Frameworks, pp. 45–50. University of Malta, Valletta, Malta (2010)
23. Rijo, R., Martinho, R., Pereira, L., Silva, C.: Text mining applied to electronic medical records. Int. J. E-Health Med. Commun. **6**(3), 1–18 (2015)
24. van Rijsbergen, C.J.: Information Retrieval, 2nd edn. Butterworth-Heinemann, Oxford (1979)
25. Rumshisky, A., et al.: Predicting early psychiatric readmission with natural language processing of narrative discharge summaries. Transl. Psychiatry **6**, e921 (2016)
26. Saito, T., Rehmsmeier, M.: The precision-recall plot is more informative than the ROC plot when evaluating binary classifiers on imbalanced datasets. PLoS ONE **10**(3), e0118432 (2015)
27. Singh, J.P., et al.: International perspectives on the practical application of violence risk assessment: a global survey of 44 countries. Int. J. Forensic Ment. Health **13**(3), 193–206 (2014)
28. Suchting, R., Green, C.E., Glazier, S.M., Lane, S.D.: A data science approach to predicting patient aggressive events in a psychiatric hospital. Psychiatry Res. **268**, 217–222 (2018)
29. Syed, S., Spruit, M.R.: Full-text or abstract? Examining topic coherence scores using latent Dirichlet allocation. In: DSAA2017, pp. 165–174 (2017)
30. Wang, Y., Wang, L., Rastegar-Mojarad, M., Moon, S., Shen, F., Afzal, N., Liu, S., Zeng, Y., Mehrabi, S., Sohn, S., Liu, H.: Clinical information extraction applications: a literature review. J. Biomed. Inform. **77**, 34–49 (2018)
31. Yin, H., Camacho, D., Tino, P., Tallón-Ballesteros, A.J., Menezes, R., Allmendinger, R. (eds.): IDEAL 2019. LNCS, vol. 11872. Springer, Cham (2019). https://doi.org/10.1007/978-3-030-33617-2

An Auto Question Answering System for Tree Hole Rescue

Fulin Wang[1(✉)] and Yun Li[1,2]

[1] School of Information Science and Technology,
Beijing Forestry University, Beijing 100083, China
`Fulin-Wang@hotmail.com, liyun@bjfu.edu.cn`
[2] Engineering Research Center for Forestry-oriented Intelligent Information Processing,
National Forestry and Grassland Administration, Beijing 100083, China

Abstract. This paper introduces an automatic question answering system which aimed to provide online how-to instructions for volunteers of Tree Hole Rescue–a Chinese online suicide rescue organization. When a volunteer needs to make sure how to deal with a rescue task professionally, he/she could ask this system via its WeChat public account other than reading a rescue instruction menu book. Firstly, a Tree Hole Rescue question-answer knowledge graph was constructed to manage Tree Hole Rescue question-answer knowledge and its relationship. Then, based on this semantic technology, a question in Chinese natural language was parsed into a machine-readable logical language through the question preprocessing and entity mapping process. And, a candidate question set was generated through a hierarchical information retrieval strategy. Finally, an exact or close answer and recommended similar questions were sent to the asker after calculating words sequence similarity via an algorithm which combined word form and semantic features. If the system could not match an answer for a question, the question would be added to unsolved question list and the system would alert administrator to deal with it. System testing shew that the Q&A system has a high accuracy rate in response of Tree Hole Rescue questions. Meanwhile, this system provides a series of methods to improve the update capability of the Q&A library and the scalability of the system.

Keywords: Tree Hole Rescue · Automatic question answering · Natural language processing · Knowledge graph · Word embedding

1 Introduction

Tree Hole Rescue Team is a non-profit organization which aims at online suicide crisis intervention. It gathers hundreds of psychology-related professional volunteers. They use intelligence robot monitoring social media and finding high-risk suicide individuals. The analysed results were sent to their Wechat groups twice a day as Tree Hole Alert Report. Members of Tree Hole Rescue Team contact and provide help to rescued objects according to those warning reports. Volunteers try to save rescuer's lives by contacting their family and friends, as well as conducting psychological counseling. In the past two

© Springer Nature Switzerland AG 2020
Z. Huang et al. (Eds.): HIS 2020, LNCS 12435, pp. 15–24, 2020.
https://doi.org/10.1007/978-3-030-61951-0_2

years, the Tree Hole Rescue Team has successfully prevented thousands of suicides and saved hundreds of lives. As its influence and scale continuing to expand, more and more volunteers join in the Tree Hole Rescue Team. And they often need find out properly how-to instructions during rescue action. The team already provides a rescue instruction manual that give clearly information to deal with kinds of situations. While, it is difficult to find out suitable solutions immediately just relying on reading manual or asking questions in Wechat groups and waiting responses. At the same time, there are many high similarity questions asked by different volunteers, which has caused a waste of human resources and affected the efficiency of rescue work. Under such circumstances, an automatic question answering system was designed and developed based on the "Online Suicide Rescue Instruction" [1] issued by Tree Hole Rescue Team. In order to manage Tree Hole Rescue question-answer knowledge and its relationship, a Tree Hole Rescue question-answer knowledge graph was constructed. The construction process was divided into three parts: Q&A data preprocessing, core concept extraction and Q&A knowledge graph construction. Then a Word2Vec model was trained to generate word embedding. During the automatic question answering task, the semantic similarity between words was calculated by word embeddings. The detailed technical solution to automatic Q&A consisted of three steps: question analysis, information retrieval and sentence similarity calculation. If the system could not match an answer for a question, the question would be added to unsolved question list and the system would alert administrator to deal with it. The automatic Q&A program was embedded into a WeChat public account's chat window. It provides convenient and intelligent services for volunteers of Tree Hole Rescue.

2 Related Work

An automatic question answering system refers to an information retrieval system that accepts questions described by users in natural language, finds corresponding answers from the knowledge base and returns them to users. The early representative English question answering systems include ELIZA [2], which is mainly for the medical field; START released by MIT [3], which is the world's first web based question answering system; AnswerBus developed by University of Michigan [4], which can support six languages; Watson launched by IBM [5], which can answer questions in various fields such as history, literature, and geography. Comparing with English, natural language processing for Chinese is more difficult because of the complexity and flexibility of Chinese expression form. The research on the Chinese automatic question answering system started late, and the representative achievements include the Auto Talk System based on the large-scale common knowledge base "Pangu" [6], and the NKI (National Knowledge Infrastructure) question answering system developed by the Chinese Academy of Sciences [7]. The automatic question answering system can be divided into different categories from different perspectives [8]. According to different fields, it can be divided into open domain-oriented automatic question answering system and professional-oriented automatic question answering system; according to the different data types of knowledge sources, it can be divided into structured data based question answering system, free text based question answering system and question-answer pairs based question answering system; according to the classification of generating feedback mechanism, it can be divided into retrieval based question answering system and generative question answering system.

The designed Tree Hole Rescue automatic question answering system is a professional-oriented retrieval question answering system. Using frequently asked question-answer pairs as the knowledge source of this system.

3 Key Implementation Methods

3.1 Construction of Tree Hole Rescue Question-Answer Knowledge Graph

Different from the traditional automatic Q&A system which is based on question-answer pairs, in this paper, a graphical conceptual model is designed. All question-answer pairs are managed properly using a Tree Hole Rescue question-answer knowledge graph. Therefore, while extracting and storing various types of questions, answers, question types, keywords, and other information, the association relationship between different question-answer pairs is mined. At the same time, the similar question-answer pairs can be effectively related together. This way of knowledge storage is taken to improve the efficiency during information retrieval process and ensure the accuracy of the candidate set.

Q&A Data Preprocessing. The main source of the Q&A data is the "Online Suicide Rescue Instruction" edited by Tree Hole Rescue Team. It's a document which contains detailed guidance for every aspect in Tree Hole Rescue. At the same time, we referred to the comment messages under Tree Hole Rescue's official Weibo account and finally sort out 230 common question-answer pairs in topic of Tree Hole Rescue. Then, by identifying interrogative words in Chinese question sentences, the Q&A data is labeled with the question category. All questions are divided into 8 categories: when (asking time), where (asking location), how (asking method or practice), what (asking definition or content), why (asking reason), who (asking the person), how much (asking the number), whether (asking yes or no). The generated Q&A dataset is used as the basic data for constructing the Tree Hole Rescue question-answer knowledge graph, which consists of 230 records and four fields: question, answer, question category and data resource

Core Concept Extraction. The first step for core concept extraction is conducting Chinese word segmentation and part-of-speech tagging on questions in Q&A dataset. Then filter out the noise words such as mood words, auxiliary words, prepositions, and non-morpheme words in the sentence according to the part-of-speech tags. Next obtain the initial keyword list after manually checking the remaining keywords.

Chinese word segmentation and part-of-speech tagging in this paper are all implemented by "Jieba" Chinese text segmentation -a Python Chinese word segmentation module. The details of the segmentation algorithm used in this module are as follows. Firstly, achieving efficient word graph scanning based on a prefix dictionary structure and building a directed acyclic graph (DAG) for all possible word combinations. Then using dynamic programming to find the most probable combination based on the word frequency. For unknown words, a HMM-based model is used with the Viterbi algorithm. Using the default part-of-speech tokenizer in Python Jieba module to tag the part-of-speech label for each word after segmentation.

After getting the initial keywords list, similar concepts are identified through the word editing distance calculation, also supplemented by manual verification. The goal of this step is to improve the efficiency of the subsequent information retrieval. For words with similar meaning, regard the highest frequency word in the original Q&A dataset as the main keyword. As a result, the thesaurus is generated and the initial keyword list is replaced by synonyms. Core concept extraction is the key construction stage of the Tree Hole Rescue question-answer knowledge graph. This stage extracts the core information in each question-answer pair, which lays out the foundation for storing and mining the association between different question-answer pairs. Finally, the original Q&A dataset is transferred into three sets: Question Set which contains 230 records and four fields of question id, question content, question category and keywords; Answer Set which contains 165 records and two fields of answer id and answer content; Q&A Set which contains 230 records and two fields of question id and answer id.

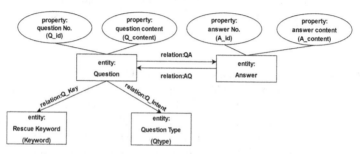

Fig. 1. The conceptual model of Tree Hole Rescue question-answer knowledge graph.

Q&A Knowledge Graph Construction. A top-down approach is adopted in the construction of the Tree Hole Rescue question-answer knowledge graph. First defining the ontology and relationship, then filling the entity. The conceptual model of the Tree Hole Rescue question-answer knowledge graph is designed as shown in Fig. 1. This model takes both the structure of current dataset and retrieval requirements of automatic question answering into consideration. After the core concept extraction stage, three sets have been generated, including question set, answer set and Q&A set. Then based on the conceptual model and these three sets to fill the entity in the knowledge graph. The open source graph database Neo4j is used to store the Tree Hole Rescue question-answer knowledge graph. In view of the limited question-answer data currently available, in the development of the system's management end, an interface for expanding the Q&A knowledge graph is reserved. Valuable question-answer pairs can be continuously added to the Q&A knowledge graph.

3.2   Word2Vec Model Training

In order to quantify and calculate the similarity between words, different words need to be expressed in numerical form. The conversion process from text to numerical value can be realized by the vector space model of words. While training the neural network, the Word2Vec tool maps the word representation from one-hot vector to a real-valued vector with a lower dimension [9], and then converts the text processing into a vector operation in the k-dimensional vector space. Through this conversion, semantic information is incorporated into the numerical representation of words. Meanwhile, the problem of dimensional disaster brought by the one-hot representation can also be solved. Each word can be represented as a unique vector after training a large number of corpora. In a vector space composed of all words, semantically similar words are located close to each other. The basic assumption of this method is that words with similar contexts will also have similar semantics [10].

Training Corpus. SogouCA, the global news data corpus provided by Sogou Lab, is used as the basic corpus. It includes news data from June to July 2012 from several news sites. The data comes from 18 channels, including domestic, international, sports, social, entertainment and so on. Text data with a size of 1.76 GB is obtained after data preprocessing. News corpus is characterized by a wide coverage, but since the main application scenarios of the Word2Vec model in this paper are related topics such as Tree Hole Rescue, depression, suicide, psychological counseling, in order to make the model cover more professional vocabulary and excavate the semantic connection of words closely related to the application scenarios, we also crawled the public news reports of Tree Hole Rescue and related articles in the field of psychology as expansion corpus. These psychology articles are from 525 psychological website [11], one psychological website [12] and Dr. Zhou depression website [13]. The total size of the expansion corpus is about 110 MB.

Training Parameters and Model Effect Test. The Word2Vec model training in this paper uses the CBOW model structure. The training parameters were set as follows: set size (the vector dimension of each word) to 400 dimensions, windows (the maximum distance between the current word and the predicted word in a sentence) to 5, alpha (learning rate) to 0.025, min_count (words whose word frequency less than it will be discarded) to 5, and the iter (number of iterations) to default value of 5. After the training is completed, a dictionary containing 425812 words is constructed. To test the training effect of the model, enter the words "Tree Hole Action" and "Tree Hole Robot" to find the top 5 vocabulary with similar semantic similarity to them respectively, and simultaneously output the similarity value. The result is shown in Fig. 2. Judging from the output results, the Word2Vec model can achieve the words' numerical quantization task accurately at the semantic level.

Fig. 2. Results for the Word2Vec model test.

3.3 Automatic Question and Answer

Question Analysis. Question parsing is aimed to analyze the sentences entered by the user and extract keywords that can be used for query. At the same time, for sentences in interrogative form, the question category is judged by Chinese interrogative words, which is also used as one of the search conditions. Since different users have various forms of natural language expressions, only by accurately understanding the user's question intention and locating the most relevant entities in the knowledge graph can the accuracy of answers be guaranteed. Figure 3 demonstrates the system's parsing process for the user's natural language questions. Firstly, preprocess the natural language question. The preprocessing process includes Chinese word segmentation, part-of-speech tagging, and noise word removal. Then, based on the pre-processed word sequence, the entity mapping operation is performed. This process only retains the vocabulary that is the keyword entity or question type entity itself, or the vocabulary that can be mapped onto the keyword entity through synonymous conversion. When judging the synonym, the first step is to judge with the thesaurus. According to the construction process of the knowledge graph, each main keyword in the thesaurus corresponds to a keyword entity in the knowledge graph. If can't get the thesaurus matching result, the second step is to use the previously trained Word2vec model to find strong related keyword entities. When the calculated similarity value between a mapped word and a keyword entity is higher than 0.5, the word will be mapped onto this entity.

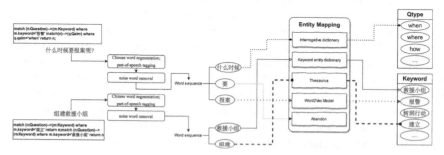

Fig. 3. Parsing process for the user's natural language questions.

Information Retrieval. In this stage, the natural language question is converted into the machine-readable logical language by constructing a Cypher query statement. The statement is generated with the keywords and question category information obtained in question analysis stage. For the two questions "When should I report the case" and "Form a rescue team", the generated query statement is shown in the boxes on the left of Fig. 3. In order to improve the retrieval efficiency while not missing out the questions that are strongly relevant to the user's question intentions, a hierarchical information retrieval strategy is designed. The first level of matching is more accurate, considering both question category and specific keywords. This part combines the keywords in the keyword list with the question category to query, and unions the queried results as the candidate set. If the first level matching fails, or the input sentence is not in a recognizable interrogative form, then ignore the question category information and enter the second layer matching. This matching level only uses keyword to perform matching, and also union the query results to obtain the candidate set. For the case where effective keywords cannot be extracted from the user's question, the user will be prompted to transfer into the manual Q&A module.

Sentence Similarity Calculation. Sentence similarity calculation within the candidate question set is the key step in this stage. The answer for specific question and recommended similar questions are finally obtained based on the calculation results. Common Chinese sentence similarity calculation methods include TF-IDF algorithm which is based on word form statistics, semantic based calculation method, as well as word-form and word-order based calculation method [14]. Among them, the TF-IDF algorithm relies on the statistical information of the word form. Its application effect is good in long texts, while the application effect in short texts similarity calculation is not ideal. The sentence similarity calculation algorithm in this paper refers to the method proposed in [15], which combines the word form and semantic features.

Firstly, the word sequences of the two sentences are obtained by word segmentation preprocessing. Then the similarity calculation is conducted from the perspective of word form and semantics respectively. Finally, the two similarities are weighted and summed. The word form similarity of two word sequences is measured by the proportion of the number of the same words in the sequence, as shown in formula (1).

$$Sim_{shape}(WS_1, WS_2) = \frac{same(WS_1, WS_2)}{len(WS_1) + len(WS_2) - same(WS_1, WS_2)} \quad (1)$$

Where WS_1 and WS_2 represent the two word sequences, $same(WS_1, WS_2)$ represents the number of the same words in the two word sequences, $len(WS_1)$ and $len(WS_2)$ represent the number of words in the two word sequences respectively, and $Sim_{shape}(WS_1, WS_2)$ is the calculated word form similarity of the two word sequences.

The calculation of word sequence similarity based on semantics is realized by the thesaurus and the pre-trained Word2Vec model. In order to increase the proportion of similarity between synonyms, the similarity of words that are synonymous with each other in the system thesaurus is set to 1, while the similarity of the remaining words is calculated according to the distance between word embedding. The calculation method is shown in formula (2)–(5).

$$Sim_{sem}(WS_1, WS_2) = \frac{\frac{Sim_1}{len(WS_1)} + \frac{Sim_2}{len(WS_2)}}{2} \quad (2)$$

$$Sim_1 = \sum_{i=1}^{len(WS_1)} max_{1 \le j \le len(WS_2)} \left| wsim\left(WS_{1i}, WS_{2j}\right) \right| \qquad (3)$$

$$Sim_2 = \sum_{j=1}^{len(WS_2)} max_{1 \le i \le len(WS_1)} \left| wsim\left(WS_{1i}, WS_{2j}\right) \right| \qquad (4)$$

$$wsim\left(WS_{1i}, WS_{2j}\right) = \begin{cases} 1, & \text{if matched in thesaurus} \\ sim_{vec}\left(WS_{1i}, WS_{2j}\right), & \text{other cases} \end{cases} \qquad (5)$$

Where $sim_{vec}\left(WS_{1i}, WS_{2j}\right)$ represents the word embedding similarity of two words, and $Sim_{sem}(WS_1, WS_2)$ is the calculated semantic similarity value between two word sequences. Finally, the word form and semantic features of the word sequence are merged, and the value of sentence similarity is calculated by formula (6).

$$Sim(WS_1, WS_2) = aSim_{shape}(WS_1, WS_2) + bSim_{sem}(WS_1, WS_2) \qquad (6)$$

Where a represents the weight of the word form similarity, and b represents the weight of the semantic similarity. According to multiple experiments, we set a to 0.3 and b to 0.7.

Finally, the questions in the candidate set are sorted according to the similarity value from high to low. The experience thresholds α and β are set to 0.6 and 0.4 respectively based on repeated experiments. If the highest similarity in the candidate set is higher than α, the answer is returned. Then if the highest similarity is lower than β, it is considered that the question cannot be retrieved. Questions whose similarity to the user's input is between α and β are considered to be strongly related to the question asked by the user, and these questions will be recommended to the user as similar questions.

4 Experiments

In view of the highly targeted application scope of the online suicide rescue Q&A, the existing open source Q&A database and knowledge base data cannot be directly used during the system development and testing process. In order to verify the accuracy of the Tree Hole Rescue question answering system, 80 natural language questions related to online suicide rescue are designed manually. The test questions cover almost all aspects of Tree Hole Rescue process. In order to make the evaluation results more convincing, the diversity of question structure is guaranteed by changing the sentence patterns and interrogative words.

The summary of test results is shown in Table 1. Among the 80 tested questions, 40 questions have found the answer, 32 questions did not directly return the answer but gave recommendations on related questions, and 8 questions could not be matched to the relevant data. For the question that directly returns the answer, if the answer matches the question, the question is considered to be answered accurately; for the case where only the recommended question is returned, if there is a match between the recommended question and the user's question content, the question is considered to be answered accurately. According to the above evaluation principles, all the 40 test questions returned with direct answer is accurately answered, while among the 32 questions returned with related questions only, 22 questions can find a correct match with at least one of the recommended questions. In this test, 62 in 80 questions are accurately answered, and the system reached an accuracy rate of 77.5%. Overall, the system has a relatively stable performance in the intelligent response task of Tree Hole Rescue questions.

Table 1. Results for the accuracy test of Tree Hole Rescue automatic question-answering.

Test result	Evaluation		Total
	True	False	
Find direct answer	40	0	40
Find related questions only	22	10	32
Return nothing	0	8	8
Total	62	18	80

Due to the currently limited Q&A data available in Tree Hole Rescue scenarios, the entity linking and thesaurus editing interface are designed to expand the knowledge range of the Q&A library. By continuously expanding the Q&A library during use, the accuracy of the system can be further improved.

5 Conclusion

In this paper, with the goal of further improving the efficiency of Tree Hole Rescue operations, an auto question answering system for Tree Hole Rescue is designed and implemented. The experiment and practical application prove that the designed scheme and technical solution are suitable for application scenarios with limited question-answer knowledge and highly targeted application groups. It shows good performance in the question answering task of Tree Hole Rescue related questions and has high practical value. However, there are still some areas for improvement in this study. For example, the automatic Q&A is based on the frequently asked question-answer pairs and question sentence matching. If the key information in the answer set can be efficiently used, more rescue knowledge may be mined through recent Q&A dataset. In addition, the current Q&A knowledge is limited in the range of "Online Suicide Rescue Instruction" document because of the lacked reference data. By introducing more authoritative materials in the area of psychology or depression, the scale of Tree Hole Rescue question-answer knowledge graph can be further expanded.

References

1. Tree Hole Rescue Team: Online Suicide Rescue Instruction. Version1.7 (2019)
2. Weizenbaum, J.: ELIZA—a computer program for the study of natural language communication between man and machine. Commun. ACM **9**(1), 36–45 (1966)

3. Katz, B., Borchardt, G.C., Felshin, S.: Natural language annotations for question answering. In: 9th International Florida Artificial Intelligence Research Society Conference DBLP, pp. 303–306 (2006)
4. Zheng, Z.: AnswerBus question answering system. In: 2nd Human Language Technology Conference, pp. 399–404 (2002)
5. Chu-Carroll, J., Fan, J.: Leveraging Wikipedia characteristics for search and candidate generation in question answering. In: 25th AAAI Conference on Artificial Intelligence, pp. 872–877 (2011)
6. Lu, R., et al.: Agent-oriented common sense knowledge base. Sci. China (Ser. E Technol. Sci.) **05**, 453–463 (2000)
7. Wang, S.: The research on QA system based on dynamic KB. In: Chinese Information Processing Society of China, pp. 597–602 (2003)
8. Zhang, Y.: Research and application of question answering technologies in traditional Chinese medicine system. Master Thesis, Zhejiang University (2018)
9. Wang, F., Tan, X.: Research on optimization strategy of training performance based on Word2Vec. Comput. Appl. Softw. **35**(01), 97–102 + 174 (2018)
10. Baroni, M., Dinu, G., Kruszewski, G., et al.: Don't count, predict! A systematic comparison of context-counting vs. context-predicting semantic vectors. In: Meeting of the Association for Computational Linguistics, pp. 238–247 (2014)
11. 525 psychological website. https://www.psy525.cn/. Accessed 01 Apr 2020
12. One psychological website. https://www.xinli001.com/. Accessed 01 Apr 2020
13. Dr. Zhou depression website. http://www.zhyyz.com/. Accessed 01 Apr 2020
14. Wang, D.: Improved sentence similarity algorithm research and its application in question answering system. Master Thesis, Dalian Jiaotong University (2010)
15. Duan, J.: Research and implementation of chinese question answering system based on frequently asked questions. Master Thesis, Xidian University (2019)

Research on the Behavior Pattern of Microblog "Tree Hole" Users with Their Temporal Characteristics

Xiaomin Jing[2,4], Shaofu Lin[1,2,4]($\boxtimes$), and Zhisheng Huang[3,4]

[1] Beijing Institute of Smart City of Beijing University of Technology, Beijing, China
linshaofu@bjut.edu.cn
[2] Faculty of Information Technology, Beijing University of Technology, Beijing 100124, China
[3] Department of Computer Science, Vrije University Amsterdam, Amsterdam, The Netherlands
[4] Advanced Innovation Center for Human Brain Protection, Capital Medical University, Beijing, China

Abstract. Depressed patients release microblog information and pay attention to each other's messages. After some depressed patients commit suicide or die for other reasons, there are still microblog messages gathering constantly, thus forming a "tree hole" - a channel for depressed patients to express their despair and suicide wish. Among them, in 2012, a microblog user died of depression, forming the largest microblog "tree hole", with more than 1.6 million messages. Based on the data of "tree hole", this paper analyzes the behavior pattern of microblog "tree hole" users according to the temporal characteristics of the message, so as to obtain the crowd behavior pattern characteristics of the potential risk persons of mental health and the potential depressed patients, it is found that the relative high incidence time of depression and suicide is related to the behavior pattern. More human and material resources can be deployed to monitor and rescue the depression suicide potential during the active period, and the results can be fed back to the relevant government departments and social rescue agencies, such as the microblog Internet police, etc., so as to make them pay attention to it and form a coordinated mode of the rescue of government-and-society.

Keywords: Depression · Microblog tree hole · Temporal characteristics · Behavior pattern · Monitor and rescue

1 Introduction

Depression is a common and easily neglected mental disease. It has become the main factor of "mental disability" in the world, especially in young and middle-aged groups. In March 2018, the World Health Organization announced that about 300 million people around the world suffer from depression, and about 800000 people commit suicide due to depression every year. Depression has greatly increased the overall burden of global diseases, which causes great suffering to patients and their families [1]. Therefore, it is urgent to carry out a systematic and in-depth study of depression and put forward effective

© Springer Nature Switzerland AG 2020
Z. Huang et al. (Eds.): HIS 2020, LNCS 12435, pp. 25–34, 2020.
https://doi.org/10.1007/978-3-030-61951-0_3

plans of monitoring and rescue. With the development of information technology, social media and so on, more and more potential depressed patients express their feelings on social media. They regard social media as a way to express themselves and a way to record their lives. These daily life tracks recorded by social media contain a lot of information of patients [2].

This paper studies the users who leave messages in the "tree hole" of microblog. They tell their thoughts and suicidal intentions in the "tree hole". Moreover, many users gather together to commit suicide. In response to this kind of suicide, the "tree hole operation rescue team" organized by a professor with the Department of Computer has been monitoring specific websites in microblog every day since July 27, 2018, finding high-risk suicide groups from microblog and issuing tree hole monitoring notification every day [3], and has rescued more than 1000 people. In this paper, according to the temporal characteristics of the messages, such as the high frequency of messages in 24 h a day and the high frequency of messages in a week. The behavior pattern of microblog "tree hole" users is analyzed according to the temporal characteristics to obtain the research results to provide more effective and accurate suggestions of monitoring and rescue, and to provide the reference for the monitoring and rescue of social institutions and the decision-making of government departments.

2 Related Work

At present, there have been many achievements in the research on depression and crowd behavior patterns at home and abroad. From the research on depression, there are some ways to solve the low recognition rate of depression. For example, through a depression recognition algorithm based on microblog text and deep learning, the existing problems of depression recognition are effectively avoided, it provides support for medical staff to actively discover and rescue patients [4]; it also analyzes the time of electronic medical records of depressed patients and messages in social media, for example, someone has studied the temporal expression in the text of electronic medical records of depression, identified the temporal expression in Chinese electronic medical records, and laid the foundation for subsequent extraction of the temporal line of electronic medical records [5]; for another example, the paper "Time Characteristics of Suicide Information in Social Media" analyzes the time characteristics based on data of microblog "tree hole", and studies the impact of holidays and major activities on depressive suicide [6].

For the study of crowd behavior patterns, both to people who suffer from a disease behavior analysis, such as, through the crowd AIDS high-risk behavior to evaluate the effect of different intervention model, provide the scientific basis for screening and take effective intervention measures [7], discusses the high-risk population health stroke the influence factors of knowledge, attitude and behavior, targeted for people at high-risk for stroke health guidance provides a theory basis [8]; In addition, social media data can be used to conduct crowd behavior analysis, so as to obtain relevant information that may help to understand specific crowd behavior and further support crowd management [9]. However, there are still few relevant studies on behavior analysis of depressed patients, and the main research is conducted on offline data, while the information of depressed patients in social media is rarely used for research.

Based on the data of online social media, this paper conducts a behavior analysis of potential patients with depression by taking "tree hole" messages on microblog. According to the characteristics of the time to leave a message, the behavior patterns of the population were analyzed to obtain the behavior characteristics of the potential mental health risk and depressed patients.

3   "Tree Hole" Data

Some users on microblog have passed away, and the page is no longer updated, but on the left page, there will be a large number of comments pouring in every day. Some people lose hope in life, and they will choose to pour out their emotions on the page where the host left, thus forming microblog "tree hole" - a channel for depressed patients to pour out. Among them, in 2012, a microblog user "Zoufan" committed suicide due to depression. Because of her indifferent attitude of leaving the world and many philosophical comments before her death, she became the idol of many depressed patients. So that her account formed the largest "tree hole" with more than 1.6 million messages so far. The research data is crawled from the "tree hole", and the research process includes data crawling, pre-processing, statistics, and visual presentation. The data in this paper rely on python programming language to write code for a long period of continuous crawling, including data of 2012 to 2019 years. The original data mainly includes comment time, microblog ID, user name, comment content, etc. In consideration of user privacy, the microblog ID and user name are not displayed. The format of data is transformed and cleaned to remove the useless duplicate data; the time is separated and corresponded to the specific week to get the preliminary processing data (Tables 1 and 2).

Table 1. The original data

Date	Microblog ID	User name	Comment content
2018-9-29 19:53	Anonymous	Anonymous	I'm tired
2018-10-1 00:18	Anonymous	Anonymous	How can I kill myself?
2018-12-6 13:05	Anonymous	Anonymous	I really can't
2019-4-12 14:18	Anonymous	Anonymous	It's no use comforting
2019-5-15 00:55	Anonymous	Anonymous	How many leaves are left on my tree

After the initial processing, the data were selected according to the condition that each user left at least 10 messages and left messages for more than 5 days. A total of 17433 qualified users and 1070052 messages were obtained. The content of message of each user is classified and counted according to the time point and week, and the behavior pattern of microblog "tree hole" users is analyzed according to the temporal characteristics of the message.

Table 2. The processed data

Date	Time	Week	Microblog ID	User name	Comment content
2018-9-29	19:53	6(Sat.)	Anonymous	Anonymous	I'm tired
2018-10-1	00:18	1(Mon.)	Anonymous	Anonymous	How can I kill myself?
2018-12-6	13:05	4(Thu.)	Anonymous	Anonymous	I really can't
2019-4-12	14:18	5(Fri.)	Anonymous	Anonymous	It's no use comforting
2019-5-15	00:55	3(Wed.)	Anonymous	Anonymous	How many leaves are left on my tree

4 The Results of Research

More than 10000 users who meet the requirements are analyzed in terms of message time characteristics. The overall message situation is shown in Fig. 1 (In this paper, the abscissa of all the analysis results related to week are from Monday to Sunday). It can be seen from the figure that most of the time for leaving messages is around the early morning and after 9 p.m. Most people start to have leisure and entertainment after 9 p.m., so the user's messages start to be active. In addition, most of the user messages are concentrated on Monday, Tuesday, and Sunday.

4.1 Divided Behavior Patterns by Messages Distributed in 24 h

Based on the time point- messages are distributed in 24 h, we divide the behavior patterns of microblog "tree hole" users into three types: " night owl", "normal time" and "no fixed mode". "Night owl" refers to users who often leave messages at night, as shown in Fig. 2. According to Baidu Encyclopedia, night owl refers to those who like to stay up late and usually stay up more than 2 am. And the late night refers to the period from midnight to dawn, about 00:00–06:00. Therefore, the time range of "night owl" mode is from 0 to 6 a.m. "Normal time" means that the message time conforms to the working and rest time of the public, as shown in Fig. 3; "no fixed mode" means that the message time is not regular and has no fixed time range, as shown in Fig. 4.

According to the statistical results at the time point, we obtained the behavior patterns and their proportions, among which the "night owl" pattern accounted for the largest proportion, 38.3%, with a total of 6,677 users. "No fixed mode" has 5,400 users; "normal time" has 5,356 users, and the results are shown in the table below (Table 3). "Night owl" accounts for more one, followed by "no fixed mode", that is, the majority of depressed potential people who leave messages actively in the late night, so it is necessary to increase human and material resources for rescue preparation in the late night, and artificial intelligence can be used for network monitoring.

4.2 Divided Behavior Patterns by Messages Distributed in the Weeks

Based on the statistical results of messages distributed in the weeks and practical application, studying the activity levels of potential suicide victims of depression on weekends

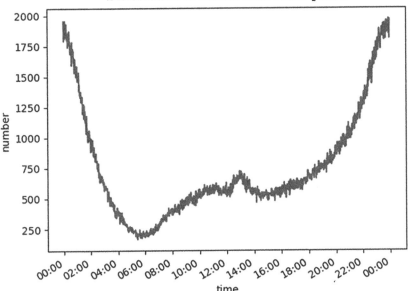

Distribution of overall active time points

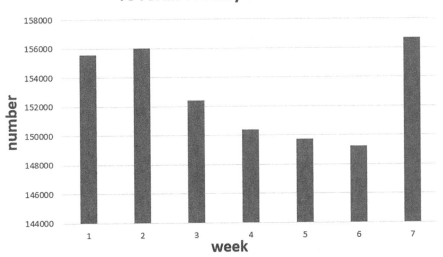

（**Overall Weekly Distribution**）

Fig. 1. The distribution of general message

could help us to carry out actions of prevention and rescue. We also divided the behavior patterns of microblog "tree hole" users into three types: "no fixed mode ", "weekend wildcat" and "workday". "No fixed mode" means that the message time is not regular and has no fixed time range, as shown in Fig. 5; "weekend wildcat" means that the message is often concentrated on the weekend, as shown in Fig. 6, "workday" means

distribution of active time points

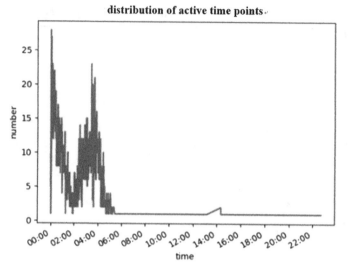

Fig. 2. Temporal distribution of "night owl" users' message

distribution of active time points

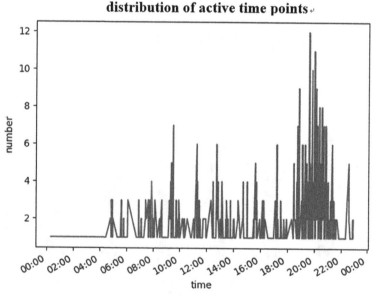

Fig. 3. Temporal distribution of "normal time" users' message

that the message is concentrated on the working days from Monday to Friday, as shown in Fig. 7.

According to the statistical results based on the week, we get the behavior patterns and their proportion, among which the "no fixed mode" accounts for the largest proportion, 50%, with a total of 8716 users; the "weekend wildcat" has 5230 users; the "workday"

ELISA

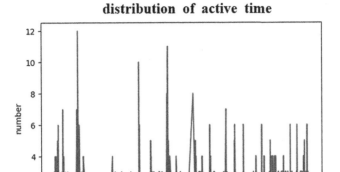

Fig. 4. Temporal distribution of "no fixed mode" users' message

Table 3. The analysis results of temporal point

Pattern	Number	Proportion
"Night owl"	6677	38.3%
"No fixed mode"	5400	31%
"Normal time"	5356	30.7%

has 3487 users, and the results are shown in the table below (Table 4). Although the proportion of "weekend wildcat" is not the largest, the number of "unfixed mode" is the largest, in this mode, the frequency of messages during weekdays is the same as that during weekends. Therefore, in a comprehensive view, the number of messages on weekends accounts for the majority, monitoring and rescue preparation on weekends cannot be ignored. We need to step up network monitoring on weekends.

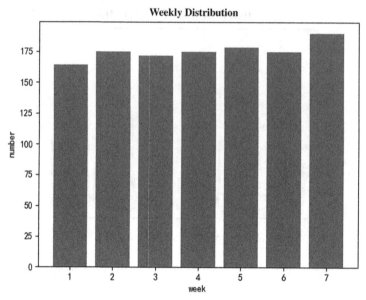

Fig. 5. Temporal distribution of "no fixed mode" users' message

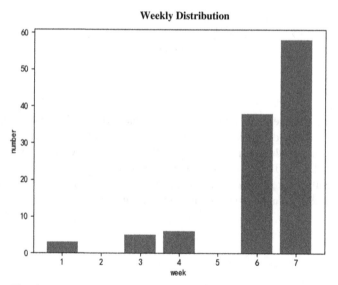

Fig. 6. Temporal distribution of "weekend wildcat" users' message

Weekly Distribution

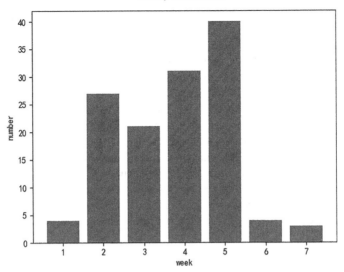

Fig. 7. Temporal distribution of "workday" users' message

Table 4. The analysis results of week

Pattern	Number	Proportion
"Weekend wildcat"	5230	30%
"Workday"	3487	20%
"No fixed mode"	8716	50%

5 Conclusion

According to the behavior patterns of microblog "tree hole" users analyzed in this paper, we draw the following conclusions:

(1) "Night owl" users have the largest number, that is, most potential depressed people leave messages in the late night. People are basically in the rest period in the late night, the intensity of artificial monitoring is not enough, so artificial intelligence can be used to monitor network message and increase the preparation of artificial rescue in the late night.

(2) According to the proportion of "no fixed mode" and "weekend wildcat", we can know that the number of weekend messages accounts for the majority, so we should pay attention to the allocation of more human resources for monitoring and rescue.

We have noticed that in social media such as microblog, some users leave messages to commit suicide. Other users will contact the provincial and municipal Internet police for checking the information and rescue. However, according to the behavior pattern of depressed patients, the high-incidence period of message and suicide is mostly in the rest period of other people, and the workers of government and social rescue such as Internet police may delay the process from finding the message to the rescue. We can feedback the results of research to social agencies of monitoring and rescue and relevant government departments, such as the "tree hole operation rescue team" and the Internet police of various provinces and cities. It is helpful to dispatch more personnel on duty at night and on rest days and allocate manpower according to the behavior pattern characteristics of the depressed people. At the same time, it can also give full play to their own characteristics and advantages, cooperate with each other, and form a coordinated mode of the rescue of government-and-society.

According to the analysis of the behavior pattern of microblog "tree hole" users based on the temporal characteristics of the message, we can understand the social activity time of the potential depressed patients, provide reference information for the treatment of depression, and arrange human resources to monitor and rescue more reasonably and effectively in time, improve the rescue rate, and rescue more suicides.

References

1. Ledford, H.: Medical research: If depression were cancer. Nature 515(7526), 182–184 (2014). https://doi.org/10.1038/515182a
2. Men, X., Wei, R., Wu, X.: Analysis and detection of language and behavior characteristics of depression in social network. J. Mod. Inf. 40(06), 76–87 2020. (in Chinese)
3. Huang, Z., Hu, Q., Gu, J., Yang, J., Feng, Y., Wang, G.: Web-based intelligent agents for suicide monitoring and early warning. J. Chin. Digit. Med. (3), 3–6 (2019)
4. Zhao, X.: Research of depression recognition based on micro-blog text and deep learning. Beijing University of Technology (2019). (in Chinese)
5. Lin, S., Zhao, Y., Huang, Z.: Time recognition of Chinese electronic medical record of depression based on conditional random field. In: Liang, P., Goel, V., Shan, C. (eds.) BI 2019. LNCS, vol. 11976, pp. 149–158. Springer, Cham (2019). https://doi.org/10.1007/978-3-030-37078-7_15
6. Huang, Z., Min, Y., Lin, F.: Time characteristics of suicide information in social media. J. Chin. Digit. Med. (3), 12–15 (2019)
7. Zhixia, L.I., Lina, W., Shaowei, J.: Analysis on influencing factors, knowledge status and behavior of AIDS among rural migrants in Weishi. Henan J. Prev. Med. (2019)
8. Yao, W.: A study on the current situation and influencing factors of health knowledge, belief and behavior of stroke high risk population. Hebei University (2019). (in Chinese)
9. Gong, V.X., Daamen, W., Bozzon, A., Hoogendoorn, S.P.: Crowd characterization for crowd management using social media data in city events. Travel. Behav. Soc. 20, 192–212 (2020)

Exploring New Opportunities for Mental Healthcare Through the Internet of Things (IoT)

Gert Folkerts[(✉)], Rogier van de Wetering, Rachelle Bosua, and Remko Helms

Open Universiteit, Valkenburgerweg 177, 6419 AT Heerlen, The Netherlands
{gert.folkerts,Rogier.vandeWetering,rachelle.bosua,
remko.helms}@ou.nl

Abstract. The Internet of Things (IoT) provide new opportunities for healthcare to enhance the quality of life and safety of mental-health patients. However, only limited studies explore how IoT can be adopted to this end, and this study tries to fill this particular gap.

This research includes a literature review and five cases with semi-constructed interviews that position the use and adoption of IoT applications in Dutch mental healthcare. This paper presents an explorative, interpretative, and mainly qualitative multiple case study. The outcomes show that privacy, security, knowledge of new technology, and opposition from traditional medical professionals are essential factors that currently reduce the acceptance of IoT in mental healthcare and lead to low IoT use.

Keywords: IoT · Internet of Things · Mental healthcare · Sensors · Privacy

1 Introduction

Are the Internet of Things (IoT) appliances at a point of a revolutionary expansion? The IoT adoption in different business contexts, particularly manufacturing, is continuing to rise [1], and there have been many developments since then. One specific area that could benefit significantly from IoT adoption is healthcare [2] because the development of healthcare systems is expanding to almost every desired level by artificial intelligence (AI) [3]. The benefit of IoT in healthcare could decrease healthcare costs, rapid access to quality care, and extending the quality of life of patients [4]. However, despite these positive consequences, the adoption of IoT in mental healthcare has only been the focus of limited research and empirical studies.

IoT applications are numerous and are still growing in number. In 2025, the number of IoT connections on the internet is predicted to be more than 100 billion [5]. IoT's applications in healthcare, also known as "internet of health things" (IoHT), are ubiquitous because healthcare lends itself exceptionally well to IoT due to the various practical options [2]. This characteristic can be seen in many applications and information technology (IT) methods for IoT-enabled healthcare [6]. In 2020, the share of healthcare-related IoT applications will even be more than 40% [7]. An analysis of IoT applications in healthcare shows that home healthcare services are the most important

© Springer Nature Switzerland AG 2020
Z. Huang et al. (Eds.): HIS 2020, LNCS 12435, pp. 35–46, 2020.
https://doi.org/10.1007/978-3-030-61951-0_4

and most significant group of IoT applications [8]. Besides that, according to Ahmadi et al. (2018), mobile, e-health, and hospital management are also essential areas that could benefit from the adoption of IoT applications within healthcare.

Mental healthcare is an integral part of National Healthcare in the Netherlands. Mental health is primarily concerned with the prevention, treatment, and the cure of mental disorders. Those with mental illnesses have to be supported through various means so that they can efficiently participate in society. When people are addicted or confused about seeking help independently, mental healthcare will assist these people. With approximately 723 mental healthcare institutions in the Netherlands, many providers offer mental health support and assistance [9]. Based on the expected positive effects of IoT in healthcare, it is likely that both patients and institutions in mental healthcare will experience positive benefits. Indications are that IoT's adoption and use in mental healthcare could improve a 'patient's quality of life and the effectiveness of his or her received medical services [10]. For example, when sensory data is used to enhance medical decision-making and enhance patient care [11]. With these positive effects of IoT adoption and use in mental healthcare, it is remarkable that limited studies about the opportunities, not to mention the threats or "hindering factors," have been conducted to point out the acceptance and implementation process of IoT in mental healthcare. From the perspective of this explorative research, it is necessary to get an answer to the following research question: *What are the main opportunities and threats associated with the adoption and use of IoT in mental healthcare?* The theoretical starting point for this explorative research is that new opportunities for IoT should add organizational value to mental healthcare institutions, benefitting both patients and practitioners. In that respect, this paper aspires to get an overview of IoT adoption and usage, specifically in the Dutch mental healthcare system.

1.1 Literature Review Method

The literature review method used in this research is a grounded literature review method in order to identify key themes related to "*IoT adoption and use in mental healthcare*". The following platforms were used to search key articles related to the topic of this study: Science Web, Google Scholar, EBSCO host, ScienceDirect (Elsevier), Science, PubMed, and the IEEE Digital Library.

1.2 Search Criteria

Empirical papers published between 1-3-2010 and 1-3-2020 with blind peer review were selected. Furthermore, journals pertaining to nursing, public health, occupational therapy and rehabilitation, psychology, social sciences, medicine, pharmacy, and therapeutical pharmacy, education, engineering, social welfare, and social work were selected. The articles were searched based on keywords such as "IoT," "mental health," "internet of things," in their title, abstract, search words, and phrases.

1.3 Literature Review Process

After finetuning and conducting a thorough search in the chosen online databases following the search criteria, a total of 70 peer-reviewed academic papers was identified.

After a filtering process, 20 academic papers were selected. Each article was read and documented in a literature review table. From this document, a layout was made based on IoT usage in mental healthcare.

2 Literature Synthesis

The literature indicates that IoT is used for different goals and purposes in mental healthcare. Based on these, the fundamental motives for adopting and using IoT can be divided into the following: 1) conducting IoT monitoring, 2) establishing control systems, 3) collecting large volumes of data (big data), and 4) performing business analytics [12]. Lee et al. (2018) based this categorization on technology trends and a performed literature review [12].

An insulin management application for diabetes with a continuous glucose monitor (CGM) is an example of an IoT device that can be used to monitor glucose level and control insulin injection. Using an IoT device, the insulin pump can be activated at a certain glucose level. This IoT system monitors and controls glucose concentration in a way the pancreas should do [13]. An example of big data and business analytics is research that profiles daily patterns of people with dementia in a Technology-Integrated Health Management (TIHM) system using IoT [11]. In this study, automated observation and sensor data through IoT are used to arrive at better clinical decisions and increase healthcare support. Based on the literature review, four critical applications associated with IoT adoption were identified. The next four sections will describe these applications: 1) Use of IoT Applications and Sensors in healthcare and 2) Use of IoT and wearables in Healthcare 3) Innovation – Use of IoT and Collaboration between IT management & Health Professionals, 4) Threats – Use of IoT and Privacy & Security. All these four sections could match one or more Lee et al. (2018) categories.

2.1 Use of IoT Applications and Sensors in Mental Healthcare

Many sensor-based applications can be used for both mental healthcare and comprehensive healthcare. For example, sensors can detect when a patient gets out of his/her bed, or sensors hidden in the floor can detect movement when somebody is out of bed, walking around. Sensors and IoT can be used to prevent social isolation with the elderly who live alone. Social isolation can cause depression or other mental health-related problems [14]. In a study performed with six older adults in the real-world, in-home experiments were carried out to see what the participants' experience was with the Internet-of-Things in terms of isolation. The method for this study drew on machine learning and sensors with IoT to monitor the activity patterns of the elderly [15]. This system is referred to as Ambient Assisted Living (AAL), a technology that facilitates independent living and keeps social connectedness between the elderly, their relatives, and medical professionals [15]. The response from the elderly regarding Forkan's study with AAL was positive. Another way of using sensors, IoT, and machine learning in daily pattern recognition is Technology Integrated Health Management (TIHM) [11]. In this method, activity patterns are used to detect changes in the daily activity of the patients. The recorded data is used to enhance the care and support given to patients and their caregiver(s). Enshaeifar's conclusion regarding the primary input of TIHM is to learn and determine variations in the patient's health and mental state. The combination of physical and mental health states are challenging issues for IoT use in mental healthcare. This combination of physical and mental health also counts for a systematic review of IoT in a new study by [16] emphasizing the trends and challenges of research on physical activity recognition and monitoring (PARM) in which IoT is an essential factor. In this review, it is believed that PARM is a crucial paradigm for smart healthcare due to the advantages of physical and mental health in rehabilitation programs. These advantages could also be visible without PARM but only monitoring the patient status with a SmartHabit system. This SmartHabit system was studied to research parameters affecting the user interaction with IT/IoT [17]. It gives the monitoring system continuous information about the status of a patient using sensors in a "smart-home setting." An important finding is that the SmartHabit system using the IoT and sensors improves the quality of care for patients who live in smart homes.

Posttraumatic stress disorder (PTSD) is a mental disorder that can appear after undergoing an awful or harrowing experience. By merging data from sensors, the IoT, and wearables in a proposed home automation system, it could be possible to predict PTSD and act in a specific way towards the patient, e.g., by suppressing nightmares slowly and waking them up quietly [18]. A health monitoring system with the combined effect of these elements could increase the quality of life for patients with PTSD.

In the major group for IoT applications, home healthcare parameters can be monitored remotely for older or sick people. Personal Healthcare Devices (PHDs) are used for this. These are portable electronic healthcare devices that record and measure the biomedical signals of users [19]. With the help of IoT technologies, PHDs can provide care professionals with supplementary information about specific characteristics of the health condition of a patient to anticipate care if needed. IoT has many benefits not only for patients but also for seniors living independently. As the elderly nowadays have longer life expectations, smart homes are a solution to allow them to live independently and for longer. Smart houses are houses in which environmental and portable medical sensors, actuators, and modern communication and IT devices are integrated [20]. These smart homes allow the elderly to live in a good home situation instead of a much more costly healthcare facility.

2.2   Use of IoT and Wearables in Mental Healthcare

A study was conducted to keep track of depression statistics and data used from GPS and smartphones in a surveillance system [21]. Poonkodi et al. (2016) used data to investigate how an unsafe moment of a patient can be detected and to keep track of the behavior of a patient and compute his/her depression level. The study of Poonkodi et al. (2016) shows that a remedying precaution can be taken to prevent negative consequences for the patient at a certain depression level. This depression level is calculated, but other systems where judgment from a health professional is needed could be possible to. Like a study with an innovative approach for a human-centered IoT model-based app that was performed. This study concerns developing a smart, responsive app that should be able to advise instantaneous activities and procedures for the patients and medical professionals. The data gathered came from IoT wearable applications. Safety protection for these medical data was a fully secured homomorphic encryption algorithm [22]. Another IoT combination with wearables is when temporal small piece patterns from patients' health are monitored. How can this be done effectively with the IoT? This is done in a case that involved remote monitoring of time-sensitive health data from students using wearables and IoT in their environment to measure and predict potential diseases based on temporal small piece pattern, for which the researchers used several grouping algorithms [23]. The developed framework was used for regaining outcomes within a specific time for medical professionals.

2.3   Opportunities - Use of IoT and Collaboration Between IT Management and Health Professionals

The collaboration between health professionals and developers/IT professionals can be improved through new IoT application development and use. The information and knowledge of these medical professionals can be embedded in the development of IoT systems and applications. [24]. Dadkhah's study also showed that there are different stages where health professionals can contribute to IoT development.

In a study, the design and development of a data integration platform with heterogeneous integrated data sources, such as data from portable or non-portable IoT devices could enable stakeholders in a patient-centered care environment to actively participate in decision-making around the patient [25].

2.4   Threats – Use of IoT and Privacy and Security

IoT data in mental health is medical data. In that case, a new environment where IoT data is applied must be well protected. The protection of IoT data also applies to the cloud. IoT requires a new security infrastructure that is based on modern technical standards. With new security design for IoT, these new standards must be taken into consideration with various attention points for security and privacy [26]. Another fact is that most IoT devices and related communication networks do not have any security procedures which make them exposed to security matters [27].

Safe data access and transport is an essential issue for medical data in the cloud. A study has been performed based on the Healthcare Industrial Internet of Things (HealthI-IoT) with ECG monitoring data and other medical data using a watermark. The use of a watermark is for extra security and safety [28]. The medical monitoring data from wearables and sensors could be sent securely and stored in the cloud. Here, health professionals can then access the medical data. When the Industrial Internet of Things (IIoT) will be enabled for continuous health monitoring, it could offer quality patient care [28]. This study is ongoing with a focus on safety and security issues.

2.5   Summary

IoT is not a single device. It is an information system that links applications like sensors, wearables to the internet, for instance, for monitoring, management, or machine learning purposes. The gathered information could, as the references indicate, be used for monitoring personal movements, health situations, machine learning, or living patterns.

The focus in the literature references is mostly on the technical content and development. Some of the references show solutions that not have been tested in practice [18, 19, 21], and the benefits given could be interpretative conclusions. References tested in practice with a small group of elderly with dementia are the TIHM [11] and SmartHabit [17]. When applications are tested like the AAL, IoT systems could use too advanced used technology, which could increase the gap between the offered and a really needed system. When the knowledge gap is too big, the acceptance of the technology by the stakeholders could be less as well. In that respect, it could be difficult to generalize "a one size fits all system" because an AAL fluctuates deeply in targeting diverse client requirements [17]. A group where the knowledge gap, as discussed with AAL, does not play a role is the TIHM. The patients do not have to decide and understand because it is done for them using the "Mental Capacity Act Code of Practice" [11]. The benefits for the TIHM could be with the health professional who has the data and the ability to judge the mental state of the patient. These benefits could play a role for the patients as well.

3   Case Study Research Method

3.1   Method

This explorative study embraces a qualitative interpretative approach. The reason why this research method was chosen is: 1) There were limited studies available about IoT use in mental healthcare, 2) The number of semi-structured interviews was low, 3) In this case, there was a need to describe and explain the main opportunities and threats associated with the adoption and use of the IoT in mental healthcare using interviews before adopting a research approach.

The research design is qualitative with an inductive research approach. This is due to the limited information in the literature and to develop a richer theoretical perspective [29]. This explorative study is based on the four themes that resulted from the literature review and five semi-structured interviews with five employees from the IT management of different mental healthcare institutions. The requirement to make a good selection for the interviewee is that the person must have decision-making powers for the IT strategy using LinkedIn search queries. The qualitative exploratory research is done using open questions with semi structured interviews based upon three subjects, IoT use, privacy and innovation [29].

3.2   Dutch Healthcare System and Case Background

The Dutch mental healthcare association has around 100 connected members, and in total, 89,000 employees work there. These employees provide care to almost one million patients per year. Every healthcare organization that deals with mental healthcare can become a member of the Dutch mental healthcare association. Among the members, there are umbrella organizations. In total, there are 1768 mental healthcare institutions in the Netherlands. Many providers in the Netherlands offer mental healthcare, and they can join the Dutch mental healthcare trade association [30]. In total, there are about 723 mental healthcare institutions [9]. 42.7% of the Dutch population between the ages of 18 and 64 have to deal with a psychological disorder during their lifetime [31] and in 2015, 1 million people were treated in a mental healthcare clinic [32].

Case size: Five organizations (representing five cases) participated in this case study and were represented by an interviewee per case; *Case I: Interviewee I (IT manager)* - an organization with 14.450 employees, 59 departments; *Case II: Interviewee II (IT manager)* - an organization with 1.333 employees, 12 departments; *Case III: Interviewee III (IT-policy manager)* - an association with 723 members – representing 86.000 employees; *Case IV: Interviewee IV (IT consultant)* - an organization with 1.306 employees, 7 departments; *Case V: Interviewee V (IT manager)* - an organization with 2.654 employees, 31 departments. These cases cover relatively 22.8% of the total employees and 15% of the entire Mental healthcare departments as part of the association. In total, there are 1723 Dutch mental healthcare institutions, including "not association members" [33].

Data Collection Method Details. As part of the case study protocol preparing data collection, procedures were taken concerning the major tasks for collecting data [34]. This included selecting organizations and interviewees, data collection procedures, case study questions, evaluations/validation of the transcribed interviews, and analysis.

Selection Criteria Interviewees. The crucial points for selecting the management of consulting roles to be interviewed were that the person must be a decision-maker who is responsible for IT/IoT adoption and implementation in mental healthcare.

Selection Organizations/Interviewees. Ultimately, five senior IT management roles from four Dutch mental healthcare institutions and one from the Dutch mental healthcare association were selected via LinkedIn and invited to participate in an interview. There would be four face to face interviews and one interview by phone. Five persons agreed to participate in the interviews: one IT policy manager, one IT consultant, and three IT managers.

Case Study Questions and Evaluations. All interviewees received a list of interview questions before attending the interview. The interviews were between 30 and 60 min in duration and recorded & transcribed before they were sent to the interviewees for validation.

Data Analysis. The data analysis proceeded with the three coding steps: 1) framework setup for coding, 2) starting initial coding, 3) defining and naming themes. Before starting with the coding, a mind map was made to code all possible connections between IoT and mental healthcare. These codes were the initial coding without seeing the transcribed interviews. The coding began with the initial coding of the transcribed interviews using NVivo and qualitative analysis. While analyzing the transcribed interviews, these codes were finetuned by adding and deleting codes along with classifying. Finally, four themes were selected based on the findings in the literature review and its match with the coding and categorization of the transcribed interviews.

4 Findings

4.1 Theme 1: Use of IoT Applications and Sensors in Mental Healthcare

Is IoT actually used in combination with sensors in the Dutch mental healthcare? From all five cases, there was only one interviewee that responded positively on the question if sensors are being used in combination with IoT. According to the interviewee's information, the main usages are sensor-based applications to detect patients coming out of bed and fall detection. When asked which IoT was used more by them in mental healthcare, one of the respondents clearly states, *"What we do further is we use sensors more often fall protection with patients. …. And what we also do is protection or a type of sensor that we can use to see if someone is still in bed." (IT manager).* Interviewee-V was using IoT to give more and more independence to the patients with their specific access to the care center. This will avoid extra workforce to support the 'patient's thanks to IoT and stimulate more patient-centered care. In that way, IoT improves the quality of life. When asking more about their IoT use, the respondent replies, *"In fact, we are increasingly endeavoring to give more control to a patient." (IT manager).* Currently, Interviewee-V is now in a pilot phase testing IoT-applications (also wearables) in cooperation with scientific institutions. New experiences learned during this pilot phase will be applied in new construction projects for mental healthcare.

4.2 Theme 2: Use of IoT and Wearables in Mental Healthcare

Wearables produce data, and Case IV is trying to work with the wearable suppliers to see if an application interface (API) could help with data integration for the EPD.

Wearables could be used to monitor the sleep rhythm and increasing the stress level of a patient. This could be of interest for mental healthcare as mentioned by a respondent: *"And we are now also doing some tests to combine that with wearables. With the idea to see whether you can also see how 'someone's sleep rhythm is based on such a wearable, but also whether you can see whether someone is increasing stress levels." (IT manager).* According to interviewee-III, with privately used wearables for health purposes, users acknowledge that privacy-sensitive medical data may be transported through the internet according, this straightforward way could be possible for medical data coming from sensors as well and used for medical monitoring.

4.3 Theme 3: Opportunities – Use of IoT and Collaboration Between IT Management and Health Professionals

To promote existing and future IoT applications, they work together with another care center in a living lab project. In this living lab, nurses, practitioners, and students can get familiarized with new technologies like sensor/IoT-applications, care robots, and apps. *"Living lab: And we hold meetings to allow nurses to become acquainted with this. So that they just get more confidence in that and a little less fear actually." (IT manager).* The 1723 members take care of their own IoT coordination. Among the five interviewees, only Interviewee-V has an official innovation steering committee. *"And from that steering group, we mainly try to get fertile ground to establish as many local initiatives as possible concerning innovative applications" .... "And we also have an Innovation committee." (IT manager).*

4.4 Theme 4: Threats – Use of IoT and Privacy and Security

IoT could be used at home or in a hospital setting. With devices that are provided by commercial parties and purchased by the patients, the Dutch mental healthcare cannot promise safe and secure medical data transportation. Also, for cloud-based applications like wearables, there is uncertainty about the GDPR. This is an essential issue at the Dutch mental healthcare from withholding using IoT. *"...of course, what do you do with the data? People must permit them to use their data. And that, of course, is something we find quite difficult as an organization." (IT manager).* Some interviewees mentioned that the main reasons the IoT is not currently used were privacy and security issues. Interviewee IV stated: *"That is the problem, originally that stuff is not yet protected in such a way." (IT Consultant).* All interviewees mention the conventional standpoint of the medical professionals, and they recognize this threshold for accepting and implementing an innovative therapy with IoT. *"No. The big problem, of course, is that a trained psychologist or psychiatrist will always assume that his observation is accurate, which electronic measuring device you can put against it" (IT manager).* & *"It is, there is a lot of resistance to innovations and things" (IT manager).*

5   Discussion and Conclusions

This study shows critical insights into the opportunities and threats perceived by senior IT managers about the adoption and use of the IoT in mental healthcare in the Netherlands. Although it is an explorative study, the five semi-structured interviews give an impression of the opportunities for the IoT use in the Dutch mental healthcare.

Only Case V uses sensors in combination with IoT devices. The gained knowledge about IoT is used to increase the mobility for patients and to act when it is necessary, for more patient-centered care [25]. The IoT technology is planned to use for a smart building project [20]. IoT and wearables are used in Case V. Data from wearables seem to be interesting in both Case IV and Case V. They are investigating how to integrate data from wearables into the EPD or just to monitor the data. By combining data from sensors and wearables into a database, the next step could come closer to IIoT [28] or TIHM [11]. Wearables seem to suffer from the unknown safety from medical data usages. The interviewees from Case V and Case IV are aware of this.

Innovation in Case IV and Case V is a cooperation between both medical professionals as well as IT professionals, not to mention the kind of innovation structures like steering committees and innovation groups. Innovation ideas are initiatives from all levels within the organization. From the five cases, only Case V was actively innovating with IoT, sensors, and wearables.

The conventional standpoint of medical professionals is mentioned a few times as a hindering factor for IoT adoption and use.

IoT is rarely used in Dutch mental healthcare. One of the reasons is safety and privacy. The interviewees cannot promise secure and safe data transportation in a home situation at the client or in a hospital setting.

5.1 Implications for Academia

The qualitative interpretations of the findings could indicate threats and opportunities for the use of IoT in mental healthcare. These findings are essential to analyze possible hindering factors that could withhold the implementation of IoT adoption. Also, it is essential for a better understanding of how innovations through IoT in mental healthcare take place with the managers, healthcare professionals, and other involved stakeholders in the IoT development and implementation process.

Research in the literature, according to the references about adoption and use in mental health care, is more focused on technology and development. The resulting benefits and additional values for the stakeholders and services are not further analyzed based on a theoretical model. Future research could be on the development of a predictive model for the organizational value of IoT within Dutch mental healthcare. Many parameters influence the adoption, use, and organizational value of IoT within mental healthcare. Some of these parameters, for example, the IoT acceptance of health professionals, should be categorized and analyzed in the mental healthcare to set up a theoretical model for the organizational value for IoT.

5.2 Implications for Practice

According to the literature review, IoT could improve the quality of life for patients within the mental health and stimulates more patient-centered care. The findings show that Dutch mental healthcare can adopt and use IoT devices like "out of bed sensors". The use of IoT innovation in Dutch mental healthcare by IoT development initiatives from within the organization in cooperation with the IT department is essential for new IoT initiatives and IoT adoption. Findings show that innovation strategy and organizational modifications such as using an innovation steering committee and innovation teams could be crucial elements to set up innovation projects for adopting and usage of IoT applications. Collaboration between IT & health professionals is part of this innovation process. During this collaboration, healthcare professionals and managers could both influence the development to gain benefits for their professional activities, patients, and the organization.

5.3 Limitations

This study is not without limitations. The findings from the respondents were not enough to extrapolate the outcomes for the whole Dutch mental healthcare. This study gives an interpretation of IoT use in mental healthcare in the Netherlands using the findings from five cases and five interviewees from five different organizations. The qualitative interpretive results from the five cases and single respondents could be useful for an organization-wide perspective follow-up research direction in the Netherlands. A quantitative evaluation of the proposed approaches was not part of this research due to the explorative, interpretative, and mainly qualitative sett-up of this multiple case study.

References

1. Mourtzis, D., Vlachou, E., Milas, N.: Industrial big data as a result of IoT adoption in manufacturing. Procedia CIRP **55**, 290–295 (2016)
2. Riazul Islam, S.M., et al.: The Internet of Things for health care: a comprehensive survey. IEEE Access **3**, 678–708 (2015)
3. Qadri, Y.A., et al.: The future of healthcare Internet of Things: a survey of emerging technologies. IEEE Commun. Surv. Tutor. **22**, 1121–1167 (2020)
4. Scarpato, N., et al.: E-health-IoT universe: a review. Management **21**(44), 46 (2017)
5. Kertiou, I., et al.: A dynamic skyline technique for a context-aware selection of the best sensors in an IoT architecture. Ad Hoc Netw. **81**, 183–196 (2018)
6. Yang, P., Xu, L.: The Internet of Things (IoT): informatics methods for IoT-enabled health care. J. Biomed. Inform. **87**, 154–156 (2018)
7. Dimitrov, D.V.: Medical internet of things and big data in healthcare. Healthc. Inform. Res. **22**(3), 156 (2016)
8. Ahmadi, H., Arji, G., Shahmoradi, L., Safdari, R., Nilashi, M., Alizadeh, M.: The application of internet of things in healthcare: a systematic literature review and classification. Univers. Access Inf. Soc. **18**(4), 837–869 (2018). https://doi.org/10.1007/s10209-018-0618-4
9. NZa: Marktscan ggz 2016 - Nederlandse Zorgautoriteit (NZa) (2016)
10. de la Torre Díez, I., Alonso, S.G., Hamrioui, S., Cruz, E.M., Nozaleda, L.M., Franco, M.A.: IoT-based services and applications for mental health in the literature. J. Med. Syst. **43**(1), 1–6 (2018). https://doi.org/10.1007/s10916-018-1130-3
11. Enshaeifar, S., et al.: Health management and pattern analysis of daily living activities of people with dementia using in-home sensors and machine learning techniques. PLoS ONE **13**(5), e0195605 (2018)
12. Lee, I., Lee, K.: The Internet of Things (IoT): applications, investments, and challenges for enterprises. Bus. Horiz. **58**(4), 431–440 (2015)
13. OPENAPS: The Open Artificial Pancreas System project (2020). https://openaps.org/
14. Lee, B., et al.: Companionship with smart home devices: The impact of social connectedness and interaction types on perceived social support and companionship in smart homes. Comput. Hum. Behav. **75**, 922–934 (2017)
15. Forkan, A., et al.: An Internet-of-Things solution to assist independent living and social connectedness in elderly. ACM Trans. Soc. Comput. **2**(4), 1–24 (2019)
16. Qi, J., et al.: Examining sensor-based physical activity recognition and monitoring for healthcare using Internet of Things: a systematic review. J. Biomed. Inform. **87**, 138–153 (2018)
17. Grguric, A., Mosmondor, M., Huljenic, D.: The SmartHabits: an intelligent privacy-aware home care assistance system. Sensors **19**(4), 907 (2019)
18. McWhorter, J., Brown, L., Khansa, L.: A wearable health monitoring system for posttraumatic stress disorder. Biol. Inspired Cogn. Arch. **22**, 44–50 (2017)
19. Park, K., Park, J., Lee, J.: An IoT system for remote monitoring of patients at home. Appl. Sci. **7**(3), 260 (2017)
20. Majumder, S., et al.: Smart homes for elderly healthcare—recent advances and research challenges. Sensors **17**(11), 2496 (2017)
21. Poonkodi, M., et al.: A comprehensive healthcare system to detect depression. Indian J. Sci. Technol. **9**(47) (2016)
22. Farooqui, M., et al.: Improving mental healthcare using a human centered internet of things model and embedding Homomorphic encryption scheme for cloud security. J. Comput. Theor. Nanosci. **16**(5–6), 1806–1812 (2019)

23. Verma, P., Sood, S.K., Kalra, S.: Cloud-centric IoT based student healthcare monitoring framework. J. Ambient. Intell. Hum. Comput. **9**(5), 1293–1309 (2017). https://doi.org/10. 1007/s12652-017-0520-6
24. Dadkhah, M., Lagzian, M., Santoro, G.: How can health professionals contribute to the Internet of Things body of knowledge. VINE J. Inf. Knowl. Manag. Syst. **49**(2), 229–240 (2019)
25. Jayaratne, M., et al.: A data integration platform for patient-centered e-healthcare and clinical decision support. Future Gener. Comput. Syst. **92**, 996–1008 (2019)
26. Li, S., Tryfonas, T., Li, H.: The Internet of Things: a security point of view. Internet Res. **26**(2), 337–359 (2016)
27. Razzaq, M.A., et al.: Security issues in the Internet of Things (IoT): a comprehensive study. Int. J. Adv. Comput. Sci. Appl. **8**(6), 383 (2017)
28. Hossain, M.S., Muhammad, G.: Cloud-assisted industrial Internet of Things (IIoT)–enabled framework for health monitoring. Comput. Netw. **101**, 192–202 (2016)
29. Saunders, M., Lewis, P., Thhornhill, A.: Research Methods for Business Students, 6th edn. Pearson Education Limited, Essex (2012)
30. GGZ. Webpage - Ggz-Sector (2019). https://www.ggznederland.nl/pagina/ggz-sector. Accessed 23 June 2019
31. De Graaf, R., Ten Have, M., van Dorsselaer, S.: De psychische gezondheid van de Nederlandse bevolking. Nemesis-2: Opzet en eerste resultaten, Trimbos-Instituut, Utrecht (2010)
32. de Ruiter, G., van Greuningen, M., Luijk, R.: Inzicht in de geestelijke gezondheidszorg, in Zorgthermometer ggz. Vektis, Enschede (2018)
33. Nederland, P.: Zorgkaart Nederland. https://www.zorgkaartnederland.nl/overzicht/organisat ietypes. Accessed 24 Mar 2019
34. Yin, K.K.: Case Study Research. Design and Methods, 3rd edn., vol. 5, p. 175 (2003). Sages Publications, Inc., Thousand Oaks

Medical Record Processing

Ontology-Based Semantic Similarity Approach for Biomedical Dataset Retrieval

Xu Wang[⊠], Zhisheng Huang, and Frank van Harmelen

Department of Computer Science, Vrije Universiteit Amsterdam,
Amsterdam, The Netherlands
{xu.wang,z.huang,Frank.van.Harmelen}@vu.nl

Abstract. Ontology-based semantic similarity approaches play an important role in text-similarity task, thanks to its ability of explanation. Ontology-based semantic similarity approaches can explain how two terms are similar with help of rich knowledge in ontology. Information retrieval aims to find relevant information for given user query. As a subareas of information retrieval, dataset retrieval is an activity to find dataset which are relevant to an information need, by using full-text indexing approach or content-based indexing approach. Ontology-based semantic similarity approaches can not only do some information retrieval tasks, such as full-text mapping, but also finding deeper similar information with the help of knowledge-richness in ontology. Because of the advantage of ontology-based similarity approaches, we are looking forwards to find the possibility to using ontology-based similarity for datasets retrieval. In this paper, we provide an ontology-based similarity approach for dataset retrieval. We run our novel approach on the bioCADDIE 2016 Dataset Retrieval Challenge. After ruining experiments, we evaluate our results with several information retrieval evaluation measures. The evaluation results show that our approach could perform well.

Keywords: Dataset retrieval · Semantic similarity · Biomedical dataset

1 Introduction

Ontology-based semantic similarity approaches are popular semantic similarity measures, which are often used to calculate similarity between two ontology concepts or terms. Semantic similarity is always classified into two types for calculating topological similarity between ontological concepts: edge-based measure and node-based measure[1].

Dataset retrieval [6] is a specialization of information retrieval, which returns a list of relevant datasets instead of documents. Dataset retrieval always considers many aspects in metadata of datasets during retrieval process, such as topic

---

[1] https://en.wikipedia.org/wiki/Semantic_similarity.

© Springer Nature Switzerland AG 2020
Z. Huang et al. (Eds.): HIS 2020, LNCS 12435, pp. 49–60, 2020.
https://doi.org/10.1007/978-3-030-61951-0_5

(vocabularies, classes, properties and individuals), time, location (geo information) and so on. However, test-content (such as title and abstract/description) of one dataset also plays an important role. For instance, if the title of one dataset shares same meaning with given query, we can also say that this dataset should be retrieved for this query.

In this paper, we present a dataset retrieval approach, which is based on ontology-based similarity measure. We choose two different type of ontology-based similarity measures to work with this dataset retrieval approach, and apply them on bioCADDIE 2016 Dataset Retrieval Challenge. With the help of gold standard in bioCADDIE 2016 Dataset Retrieval Challenge, we use both average precision (AP) [14] and normalized Discounted Cumulative Gain (nDCG) [4] measure to evaluate our approaches. Finally, we find that ontology-based semantic similarity approaches could perform well on biomedical dataset retrieval.

The main contributions of this paper are: 1) we provide two novel ontology-based semantic similarity approaches for biomedical dataset retrieval; 2) we set up an experiment to evaluate the dataset retrieval approach with the help of bioCADDIE 2016 Dataset Retrieval Challenge; 3) we find that ontology-based semantic similarity approaches could perform well on biomedical dataset retrieval task.

2 Preliminaries

2.1 Ontology-Based Semantic Similarity

As we introduced above, semantic similarity are categorised into two type for calculating similarity between ontological concepts: edge-based measure and node-based measure. Edge-based measure often consider edge counting in ontology hierarchical structure. There are many existing edge-based measures, such as Wu-Palmer [13], Pekar [10], Cheng and Cline [1] and so on. Node-based measure will consider nodes and their properties in ontology, where nodes always means the concepts in ontology. Many existing and popular node-based measures could be used, such as Resnik [12], Lin [7], Align Disambiguate Walk (ADW) [11], Jiang and Conrath [5] and so on. In this section we will introduce Wu-Palmer measure and Resnik measure, which we used in our experiments.

Wu-Palmer. Wu-Palmer measure is a very popular edge-based semantic similarity measure. It has some advantage: 1) normalized measure; 2) similarity score never goes zero; 3) no distinction between similarity/relatedness. Only thing to notice is that Wu-Palmer heavily rely on quality of graph. So, with help of a good quality ontology, Wu-Palmer could work very well for similarity task. Least common subsumer is a very important terminology in ontology-based semantic similarity measure, which is used not only in Wu-Palmer measure but also in Resnik measure. In the hierarchical structure of ontology, least common subsumer of two nodes is the lowest node which has these two nodes as descendants. Given two concepts $C1$ and $C2$ in ontology, Wu-Palmer similarity between them is:

$$Sim_{WP} = \frac{2 * Dep(LCS(C1, C2))}{Dep(C1) + Dep(C2))} \tag{1}$$

where $LCS(C1, C2)$ is the least common subsumer of $C1$ and $C2$; $Dep(C1)$ is the least path from ROOT node to $C1$.

Resnik. Resnik measure is a semantic similarity measure based on the information content (IC) of the least common subsumer. The information content of a concept is the logarithm of the probability of finding the concept in a given corpus. We chose to use Resnik measure because many other node-based similarity measures, such as Lin and Jiang$Conrath, are based on Resnik. Given two concepts $C1$ and $C2$ in ontology, Resnik similarity between them is:

$$Sim_{Resnik} = -\log P(LCS(C1, C2)) = -\log \frac{\sum_{c \in Words(LCS(C1,C2))} count(c)}{N} \tag{2}$$

where $Words(LCS(C1, C2))$ is the set of instances (i.e. terms) of $LCS(C1, C2)$; $count(c)$ is the number of counting all of c in $Words(LCS(C1, C2))$; N is the whole number of instances in ontology.

Similarity Between Sets of Concepts. After introducing two measure for calculating similarity between two concepts, we will introduce similarity measure for two sets of concepts. Given two sets $S1$ and $S2$ of concepts, the similarity between $S1$ and $S2$ is:

$$Sim(S1, S2) = \frac{sum\{Sim(C1, C2)|C1 \in S1, C2 \in S2\}}{|S1| * |S2|} \tag{3}$$

where $Sim(C1, C2)$ is the similarity between concepts $C1$ and $C2$ by Wu-Palmer or Resnik; $|S1|$ and $|S2|$ is the size of sets $S1$ and $S2$, respectively.

2.2   Evaluation Measure for Retrieval Approach

We use evaluation measures of information retrieval to evaluate our dataset retrieval approaches. In this paper, we use two popular evaluation measures, which are average precision (AP) and normalized discounted cumulative gain (nDCG), to evaluate our experiment results.

Average Precision. Average precision (AP) is a measure for ranked retrieval results [14]. For one information need, the average precision is the mean of the precision scores after each relevant document is retrieved. Average precision at position n is defiend as follow:

$$AP_n = \frac{\sum_{k=1}^{n} P@k}{R} \tag{4}$$

where $P@k$ is the precision of the top-k retrieved documents; R is the number of all retrieved datasets.

Normalized Discounted Cumulative Gain. Discounted cumulative gain (DCG) is a measure of ranking quality, which uses a graded relevance list of documents/datasets from the result to evaluate the gain of a document/dataset based on its position in the result ranking list [3]. Ideal discounted cumulative gain (IDCG) is the DCG score of ideal ranking list (list ranked by graded relevance). Normalized discounted cumulative gain (nDCG) is the normalized measure based on DCG [4]. DCG, IDCG and nDCG through top rank position p are defined as follow:

$$DCG_p = \sum_{i=1}^{p} \frac{2^{rel_i} - 1}{\log_2(i + 1)}, \tag{5}$$

$$IDCG_p = \sum_{i=1}^{|Rel_p|} \frac{2^{rel_i} - 1}{\log_2(i + 1)}, \tag{6}$$

$$nDCG_p = \frac{DCG_p}{IDCG_p} \tag{7}$$

where rel_i is the graded relevance score in position i.

3   Case Study: bioCADDIE Dataset

We use bioCADDIE Dataset 2016 as a case study of the proposed approach. The main objective of the 2016 bioCADDIE Dataset Retrieval Challenge is to create innovative ways for biomedical researchers to search and discover biomedical research data[2]. For the challenge, a collection of metadata (structured and unstructured) from biomedical datasets were generated from a set of 20 individual repositories. It consist of about 800,000 datasets of metadata. The following is an example of a metadata description with the DATS model in JSON. In this example, we can see that the property "DOCNO" is used to define the local ID of a dataset. The "DOCNO" follows by the property "metadata" which states various values of the data, which includes the data origin such as "clinical_study", data item such as data type, study group, grant, detailed study description, dataset identifier, detailed dataset description (title, description, and others).

```
{"DOCNO": "10", "METADATA":
 {"dataResource":
  {"keywords": [], "altNames": [], "acronyms": []},
     "citation": {"count": "0"},
     "organism": {"experiment": {"species": "Homo sapiens"}},
     "dataItem": {"description": "gene expression at 6h of
differentiation of Human endometrial stromal cell
expressing either or both of PRA and PRB Endogenous PGR
expression is silenced with siRNA mediated knockdown.
Then, cells are transduced with adenovirus expressing flag
```

---

[2] https://biocaddie.org/biocaddie-2016-dataset-retrieval-challenge-registration.

```
tagged PRA or flag tagged PRB. Human endometrial
stromal cell expressing one or both isoforms are treated
with differentiation cocktail for 6h.",
  "title": "Expression data of PGR isoforms, PRA and PRB,
  regulated genes in differentiating human endometrial
  stromal cells",
  "releaseDate": "2015-04-27","lastUpdateDate": "2015-05-02",
  "dataTypes": ["organism","dataItem","citation"],
  "ID": "522705",
  "experimentType": "transcription profiling by array"}},
  "REPOSITORY": "arrayexpress_020916",
  "TITLE": "Expression data of PGR isoforms, PRA and PRB,
  regulated genes in differentiating human endometrial
  stromal cells"}
,...
```

The main advantage of using bioCADDIE Dataset 2016 is that the organizers provide a set of test queries and their expected answers, which can serve as a gold standard for the evaluation of our proposed approach.

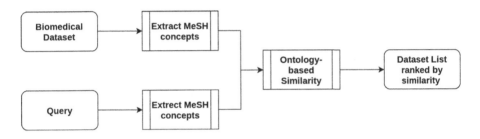

Fig. 1. Pipeline of similarity-based biomedical dataset retrieval.

4   Similarity-Based Biomedical Dataset Retrieval

In this section we will introduce the pipeline of using ontology-based semantic similarity for biomedical dataset retrieval, which is shown in Fig. 1.

Because of the gold standard of 2016 bioCADDIE Dataset Retrieval Challenge, our biomedical dataset retrieval approach aims to find a ranking list of datasets for each query. So we will introduce a whole end2end process (from given queries and datasets to final ranking list) in this section.

4.1   Biomedical Concepts Extraction

We using ontology entity exact-matching approach to extract biomedical entities from datasets, and return concepts which include extracted entities in ontology. Given a dataset and an ontology, Exact-matching extracts words/phrases which are not only words/phrases in dataset but also entities in ontology.

Definition 1. *Given dataset D and ontology O, exact-matching would extract a set of matched entities $EM_O(D) = \{e|e \in W(D) \cap E(O)\}$, where $W(D)$ is the set of all words/phrases of D and $E(O)$ is the set of all entities in O.*

For exact-matching approach, we also have a proposition. Exact-matching would not cover any words/phrases which are not ontology entity, but can not cover all the ontology entities in datasets. This is because exact-matching approach is soundness but not completeness.

Proposition 1. *Given dataset D, ontology D and a set of matched entities $EM_O(D)$. Then we have:*

- $W(D) \models e^+, \forall e^+ \in EM_O(D)$
- $E(O) \not\models e^-, \forall e^- \notin EM_O(D)$.

Example 1. We show an example of bioCADDIE dataset and the results of exact-matching for this dataset. The example dataset is formatted as JSON as follow:

```
{
"id":"793860",
"abstract":"The goal of the study is to examine how cognitive behavioral therapy (
    CBT), a common treatment for adolescent depression, affects brain functioning.
    Depressed adolescents will complete an initial assessment consisting of an
    interview, questionnaires, computer tasks, and an EEG recording. EEG (
    electroencephalography) measures brain activity by recording the electrical
    activity along the scalp caused by the firing of neurons within the brain. They
    will then complete 12 sessions of cognitive behavioral therapy, which will be
    50 minutes long and held once a week for 12 weeks. Before their third therapy
    session, participants will complete a computer task while EEG data are recorded
    . After completing the treatment, the participants will complete a final
    assessment, which will include questionnaires, computer tasks, and an EEG
    recording. They will also complete follow-up assessments over the phone 1, 3,
    and 6 months after completing the treatment. This study will also include
    healthy control participants. They will complete an initial assessment
    consisting of an interview, questionnaires, computer tasks, and an EEG
    recording. Three weeks later, they will return to complete a behavioral task
    while EEG data are recorded. Twelve weeks after the initial assessment, they
    will complete a final assessment, which will include questionnaires, computer
    tasks, and an EEG recording.",
"title":"Examination of the Neural Components Underlying the Treatment of Adolescent
    Major Depression"
}
```

Then we use exact-matching approach to extract entities from title and abstract, and return concepts which include extracted entities in ontology. In MeSH ontology, we consider MeSH Descriptor as our concept in general. MeSH Descriptor consists of a set of MeSH Concepts and a set of MeSH Terms[3]. In one MeSH Descriptor, all the MeSH Concepts are synonymous with each other. All the MeSH Terms in same MeSH Descriptor are the entities of the MeSH Concepts in this MeSH Descriptor. The extracted concepts (matched entities and MeSH Descriptor IDs in brackets) are shown as follow:

```
{
"Extracted_Concepts_from_Title":["Depression(Depression,D003863)","Adolescent(
    Adolescent,D000293)"],
"Extracted_Concepts_from_Abstract":["Goals(Goal,D006040)","Cognitive Behavioral
    Therapy(Cognitive Behavioral Therapy,D015928)","Depression(Depression,D003863)
    ","Brain(Brain,D001921)","Adolescent(Adolescent,D000293)","Surveys and
    Questionnaires(Questionnaires,D011795)","Computers(Computers,D003201)","Weights
    and Measures(Measures,D014894)","Scalp(Scalp,D012535)","Neurons(Neurons,
    D009474)","Electroencephalography(EEG,D004569)"]
}
```

---

[3] https://www.nlm.nih.gov/mesh/concept_structure.html.

4.2 Biomedical Concepts Similarity with MeSH

In MeSH ontology, each MeSH Descriptors have at least one tree number in MeSH Tree Structure[4]. With the given tree number, we can easily use both Wu-Pamler and Resnik similarity to calculate similarity between MeSH Descriptors. This is because we can find the least common subsumer of two MeSH Descriptors by using tree number. For instance, in MeSH ontology, Goals (D006040) has tree number "F01.658.500" and Depression (D003863) has tree number "F01.145.126.350". Then we can easily see that Goals and Depression has lease common subsumer with tree number "F01", which is Behavior and Behavior Mechanisms (D001520). Also with the help of tree number in MeSH ontology, we can easily know the shortest path from ROOT to MeSH Descriptor. We continue with above example of Goals and Depression. Tree number "F01.658.500" means the shortest path from ROOT of MeSH to Goals is 3. Tree number "F01.145.126.350" means the shortest path from ROOT of MeSH to Depression is 4. Tree number "F01" states that Behavior and Behavior Mechanisms, the least common subsumer of Goals and Depression, only has shortest path 1 to ROOT of MeSH. Then we have Wu-Palmer similarity between Goals and Depression, which is $\frac{2*1}{3+4} = 0.2857$. For Resnik similarity, we try to find all the MeSH Descriptors, which have "F01" in their tree number. Then we count the number of all the MeSH terms in these MeSH Descriptors, which is 1735. Also, we count all the MeSH terms in MeSH ontology, which is 284832. Now we have Resnik similarity between Goals and Depression, which is $- \log \frac{1735}{284832} = 2,2152$.

4.3 Similarity-Based Ranking List

In 2016 bioCADDIE Dataset Retrieval Challenge, the gold standard is a set of test queries and their expected answers (a list of datasets ranked by relevance score for each query). So our final results are lists of datasets, one list of datasets per query. After we calculate all the Wu-Palmer and Resnik similarity, we could rank all the dataset candidates based on similarity results for each query.

5 Experiments

In this section we will introduce our experiments of using ontology-based similarity approach for bioCADDIE Dataset Retrieval Challenge.

5.1 Biomedical Datasets Retrieval Experiment

As we mentioned in previous section, our biomedical datasets retrieval approach is based on Wu-Palmer and Resnik similarity measures. The output of experiments are ranked list of datasets based on similarity results.

---

[4] https://www.nlm.nih.gov/mesh/intro_trees.html.

Datasets. The datasets we used are all from 2016 bioCADDIE Dataset Retrieval Challenge. We use these datasets because they are all biomedical datasets from bioCADDIE, a very popular biomedical dataset search engine. Also, all the datasets include text-content in their metadata, such as title and abstract. We can use our entity matching approach and ontology-based similarity measures on these text-content directly.

Query. 15 queries we used in our experiments are from 2016 bioCADDIE Dataset Retrieval Challenge:

1. Find protein sequencing data related to bacterial chemotaxis across all databases
2. Search for data of all types related to MIP-2 gene related to biliary atresia across all databases
3. Search for all data types related to gene TP53INP1 in relation to p53 activation across all databases
4. Find all data types related to inflammation during oxidative stress in human hepatic cells across all databases
5. Search for gene expression and genetic deletion data that mention CD69 in memory augmentation studies across all databases
6. Search for data of all types related to the LDLR gene related to cardiovascular disease across all databases
7. Search for gene expression datasets on photo transduction and regulation of calcium in blind D. melanogaster
8. Search for proteomic data related to regulation of calcium in blind D. melanogaster
9. Search for data of all types related to the ob gene in obese M. musculus across all databases
10. Search for data of all types related to energy metabolism in obese M. musculus
11. Search for all data for the HTT gene related to Huntington's disease across all databases
12. Search for data on neural brain tissue in transgenic mice related to Huntington's disease
13. Search for all data on the SNCA gene related to Parkinson's disease across all databases
14. Search for data on nerve cells in the substantia nigra in mice across all databases
15. Find data on the NF-kB signaling pathway in MG (Myasthenia gravis) patients

Gold Standard. As we mentioned in Sect. 3, we use given a set of queries and their expected answers as gold standard. For each query, its expected answers are a list of bioCADDIE datasets ranked by relevant score. The relevant score are ranged from 0 to 3:

- score 0: no relevance between dataset and query;
- score 1: possible no relevance between dataset and query;
- score 2: possible relevance between dataset and query;
- score 3: relevance between dataset and query.

Evaluation. As we mentioned above, we would use AP and nDCG measure to evaluate our retrieval approach. The relevant score (ranged from 0 to 3) in gold standard could be used in nDCG directly. This is because relevant score ranged from 0 to 3 is also used as standard in definition of nDCG [3]. However, for average precision (AP) measure, the relevant score is always ranged from 0 (not relevance) to 1 (relevance). So split our AP evaluation experiments into two scenarios:

- Brave scenario: We consider score 2 and 3 of relevant score ranged from 0 to 3, as score 1 of relevant score ranged from 0 to 1.
- Caution scenario: We consider only score 3 of relevant score ranged from 0 to 3, as score 1 of relevant score ranged from 0 to 1.

5.2   Results

Now we can go straight to results of evaluation for our dataset retrieval approach. In Table 1, we can see the evaluation results for dataset retrieval approach based on Wu-Palmer. For all the scenarios of AP measure, the evaluation results are not good. In Chapter 8 of [8], the disadvantage of AP measure is introduced: AP is the least stable of the commonly used evaluation measures and that it does not average well, since the total number of relevant documents for a query has a strong influence on precision at k. So we can know the reason why the results of AP measure are not stable. For nDCG evaluation scores, we can find that nDCG scores are quite good while nDCG@10 scores are not good. This means that our dataset retrieval approach performs not so well if we just consider Top10 retrieved datasets. The highest nDCG score can reach almost 0.9 (0.8928), for Query 15. This is a very good result for nDCG evaluation. For this query, nDCG@10 score is 0.6976, a very good score in nDCG@k evaluation. We also achieve high nDCG score in Query 1,3,4,6,14. This means that in these queries, our approach can retrieve very relevant datasets by using edge-based similarity measure. A more clear and direct view can be found in Fig. 2.

Table 1. Evaluation results of dataset retrieval approach based on Wu-Palmer.

QueryID	Brave_AP	Brave_AP@10	Caution_AP	Caution_AP@10	nDCG	nDCG@10
1	0.2674	0.9091	0.0639	0.4826	0.6813	0.5566
2	0.0011	0	0	0	0.2963	0
3	0.2857	0.3754	0.0002	0	0.7867	0.2543
4	0.0438	0	0.0002	0	0.5321	0
5	0.0014	0	0	0	0.2914	0
6	0.119	0.1192	0.0006	0	0.6749	0.1004
7	0.0027	0	0	0	0.4668	0
8	0.0022	0.0227	0	0	0.3668	0.0489
9	0.0251	0	0.0029	0	0.4376	0
10	0.0551	0.0283	0.0054	0	0.4777	0.0433
11	0.0135	0	0.0083	0	0.4	0
12	0.0422	0.0909	0.0039	0	0.4433	0.0734
13	0.0134	0	0.0025	0	0.3945	0
14	0.0565	0.1111	0.0267	0.1111	0.4756	0.1909
15	0.2129	0.4527	0	0	0.8928	0.6976

From Table 2 and Fig. 3, we can state that dataset retrieval approach based Resnik has evaluation results similar to Wu-Palmer one. Resnik-based approach also has poor scores in all AP scenarios but have good scores in nDCG scenario. The highest nDCG score is 0.8888 (Query No. 15). For this query, the nDCG@10 score can also reach 0.6563. So we can know that our approach performs very well in this query, according to nDCG evaluation. We also achieve high nDCG score in Query 1, 3, 4, 6, 10, 12, 14. This means our approach can work well in some queries with both edge-based and node-based similarity measures.

Summary of Results. Through the evaluation results, we can have a brief summary for our approach's performance:

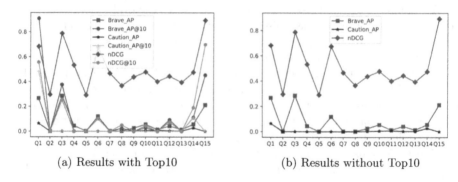

(a) Results with Top10   (b) Results without Top10

Fig. 2. Evaluation results for dataset retrieval approach based on Wu-Palmer.

Table 2. Evaluation results of dataset retrieval approach based on Resnik.

QueryID	Brave_AP	Brave_AP@10	Caution_AP	Caution_AP@10	nDCG	nDCG@10
1	0.2619	0.8182	0.0618	0.5071	0.6813	0.5535
2	0.0013	0	0	0	0.3142	0
3	0.2857	0.3754	0.0002	0	0.7867	0.2543
4	0.0497	0	0.0003	0	0.5435	0
5	0.0014	0	0	0	0.2971	0
6	0.1412	0.6032	0.0004	0	0.7178	0.2628
7	0.0037	0.0455	0	0	0.5214	0.1389
8	0.0022	0	0.0001	0	0.3662	0
9	0.0251	0	0.0029	0	0.4376	0
10	0.0753	0.1662	0.0056	0	0.5127	0.1223
11	0.0214	0.053	0.0135	0.053	0.4534	0.1155
12	0.0873	0.2489	0.0137	0.1169	0.5697	0.2585
13	0.0171	0.0227	0.0026	0	0.4106	0.0316
14	0.0713	0.1545	0.0371	0.1545	0.5116	0.2459
15	0.216	0.4057	0	0	0.8888	0.6563

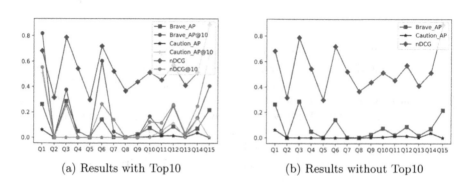

(a) Results with Top10   (b) Results without Top10

Fig. 3. Evaluation results for dataset retrieval approach based on Resnik.

- Our approach performs well in nDCG evaluation (very well in some queries, such as Query 15);
- Our approach cannot get a good result if we just consider Top10 retrieved datasets;
- Our approach performs not well in AP evaluation, and we discuss the possible reason might be the disadvantage of AP evaluation.

6   Conclusion and Discussion

In this paper, we provided a new dataset retrieval approach based on ontology-based semantic similarity. We test this approach on 2016 bioCADDIE Dataset Retrieval Challenge, and evaluate it with average precision (AP) measure and normalized discounted cumulative gain (nDCG) measure. The evaluation result shows that our approach can perform well in dataset retrieval task.

We also have some future work to do. Other similarity measures of calculating similarity between test-content can be used, such as Word2vec [9] and normalized Google Distance (NGD) [2]. In this paper we also find some disadvantage of our approach, which would be improved in future work.

Acknowledgments. This work has been funded by the Netherlands Science Foundation NWO grant nr. 652.001.002, it is co-funded by Elsevier B.V., with funding for the first author by the China Scholarship Council (CSC) grant number 201807730060.

References

1. Cheng, J., et al.: A knowledge-based clustering algorithm driven by gene ontology. J. Biopharm. Stat. **14**(3), 687–700 (2004). https://doi.org/10.1081/BIP-200025659
2. Cilibrasi, R.L., Vitanyi, P.M.: The google similarity distance. IEEE Trans. Knowl. Data Eng. **19**(3), 370–383 (2007)
3. Järvelin, K., Kekäläinen, J.: IR evaluation methods for retrieving highly relevant documents. In: Proceedings of the 23rd SIGIR Conference, SIGIR 2000, pp. 41–48. ACM, New York (2000)
4. Järvelin, K., Kekäläinen, J.: Cumulated gain-based evaluation of IR techniques. ACM Trans. Inf. Syst. **20**(4), 422–446 (2002). https://doi.org/10.1145/582415. 582418
5. Jiang, J.J., Conrath, D.W.: Semantic similarity based on corpus statistics and lexical taxonomy. In: Proceedings of the 10th Research on Computational Linguistics International Conference. The Association for Computational Linguistics and Chinese Language Processing (ACLCLP), Taipei, Taiwan, pp. 19–33, August 1997. https://www.aclweb.org/anthology/O97-1002
6. Kunze, S.R., Auer, S.: Dataset retrieval. In: 2013 IEEE Seventh International Conference on Semantic Computing, pp. 1–8 (2013)
7. Lin, D.: An information-theoretic definition of similarity. In: Proceedings of the Fifteenth International Conference on Machine Learning, ICML 1998, pp. 296–304. Morgan Kaufmann Publishers Inc., San Francisco (1998)
8. Manning, C.D., Raghavan, P., Schütze, H.: Introduction to Information Retrieval. Cambridge University Press, Cambridge (2008)

9. Mikolov, T., Corrado, G., Chen, K., Dean, J.: Efficient estimation of word representations in vector space, pp. 1–12, January 2013
10. Pekar, V., Staab, S.: Taxonomy learning: factoring the structure of a taxonomy into a semantic classification decision. In: Proceedings of the 19th International Conference on Computational Linguistics, COLING 2002, vol. 1, pp. 1–7. Association for Computational Linguistics, USA (2002). https://doi.org/10.3115/1072228.1072318
11. Pilehvar, M.T., Jurgens, D., Navigli, R.: Align, disambiguate and walk: a unified approach for measuring semantic similarity. In: Proceedings of the 51st Annual Meeting of the Association for Computational Linguistics (Volume 1: Long Papers), Sofia, Bulgaria, pp. 1341–1351. Association for Computational Linguistics, August 2013. https://www.aclweb.org/anthology/P13-1132
12. Resnik, P.: Using information content to evaluate semantic similarity in a taxonomy. In: Proceedings of the 14th International Joint Conference on Artificial Intelligence, IJCAI 1995, vol. 1, pp. 448–453. Morgan Kaufmann Publishers Inc., San Francisco (1995)
13. Wu, Z., Palmer, M.: Verbs semantics and lexical selection. In: Proceedings of the 32nd Annual Meeting on Association for Computational Linguistics, pp. 133–138. Association for Computational Linguistics (1994)
14. Zhang, E., Zhang, Y.: Average Precision, pp. 192–193. Springer, Boston (2009). https://doi.org/10.1007/978-0-387-39940-9_482

Research on Named Entity Recognition of Traditional Chinese Medicine Electronic Medical Records

Feng Lin and Dan Xie$^{(\boxtimes)}$ (iD)

Hubei University of Chinese Medicine, Wuhan 430065, China
dinaxie@hbtcm.edu.cn

Abstract. The electronic medical record (EMR) is a patient's individual medical record written by health care providers to describe the medical activities of patients. Named entity recognition (NER) of EMR is helpful to extract important information from a large number of unstructured texts, which lays a foundation for medical data mining and application. The named entity of Traditional Chinese Medicine (TCM) is more complex and its length is uncertain. In order to explore the effective method of named entity recognition of TCM EMR, after comparing the existing entity recognition methods and models, this paper selects three models, BiLSTM-CRF, lattice LSTM-CRF and BERT, to recognize the symptom entities in EMR, and carries out comparative experiments. After the real EMR data was manually labeled, three models were used to train, and the precision, recall and F1 value were used to evaluate the recognition effect of the model. The experimental results show that BERT model has the best recognition effect about TCM EMR, the precision is 89.94%, the recall is 88.27%, and the F1 value is 89.10%.

Keywords: Named entity recognition · Traditional Chinese Medicine Electronic Medical Records · BiLSTM-CRF · Lattice LSTM-CRF · BERT

1 Introduction

Electronic medical record is a record written by health care providers to describe patients' medical activities, including a series of important information closely related to patients' health status, such as patients' diseases, symptoms and signs, examination and treatment. Massive EMR data contains a large amount of medical knowledge. It can promote the research and utilization of EMR by automatically transforming unstructured EMR data into computer recognizable structured data [1].

Named entity recognition refers to the recognition of entities with specific meaning in text, such as person name and place name. Different from the named entities in the general domain, EMR contains entities such as diseases and symptoms. Named entity recognition plays a key role in the structure of text information. Named entity recognition technology can be divided into three categories: (1) rule-based and dictionary-based methods, which need domain experts to write rules and dictionaries. (2) statistical model-based methods,

© Springer Nature Switzerland AG 2020
Z. Huang et al. (Eds.): HIS 2020, LNCS 12435, pp. 61–67, 2020.
https://doi.org/10.1007/978-3-030-61951-0_6

mainly including hidden Markov models (HMM), conditional random fields (CRF) and so on, which depend on the correctness of feature selection; (3) Machine learning-based methods. In recent years, it has gradually become the mainstream method. The advantage of this method is that it can automatically extract features from the established neural network model, which reduces the workload of artificial feature formulation. In 2016, Lample et al. [2] adopted the model based on BiLSTM and CRF to obtain the best performance of NER at that time in four languages. The BiLSTM model, which takes full account of the context information of long-distance time series, is the current mainstream named entity recognition model.

The naming entity of traditional Chinese medicine has complex characteristics and uncertain length. In addition, there are sub entities in the entity. Therefore, the work of named entity recognition in traditional Chinese medicine is more complex and more difficult than that in general fields [3]. At present, BiLSTM-CRF model is the most widely used technology in TCM named entity recognition. Li Minghao et al. [4] proposed a recognition method based on long-term and short-term memory networks and conditional random fields for clinical symptom terms of TCM cases, with the highest F1 value of 0.78. Gao su et al. [5] used BiLSTM-CRF model to identify five entities in Huangdi Neijing, such as TCM cognition method, TCM Physiology, TCM Pathology, TCM nature, treatment principle and treatment method, with the accuracy rate of 85.44%, recall rate of 85.19% and F1 value of 85.32%. Zhang Yipin et al. [6] used traditional Chinese Medicine Classics as data source and BiLSTM-CRF model to extract entities such as diseases, prescriptions and herbal medicines of traditional Chinese medicine.

In 2018, Yue Zhang et al. [7] proposed a lattice LSTM model for Chinese named entity recognition. Experiments on multiple datasets show that lattice LSTM is superior to the word based and character based LSTM baseline models. In October 2018, the performance of the Bert model proposed by Devlin et al. [8] on 11 NLP tasks refreshed the record, including named entity recognition tasks. Therefore, three models, BiLSTM-CRF, lattice LSTM-CRF and BERT, are selected for experiments to determine the model suitable for named entity recognition of TCM EMR.

2 Model

2.1 BiLSTM-CRF

The structure of BiLSTM-CRF [2] is shown in Fig. 1, which is mainly divided into three layers. The first layer is the word vector layer, which maps the input words into vectors. The second layer is BiLSTM layer, which is composed of forward LSTM, backward LSTM and output layer. The last layer is CRF layer, which can predict the current input state through the past input and the state of the input, and ensure that the tag sequence is not only the most probability but also the most consistent with the language order.

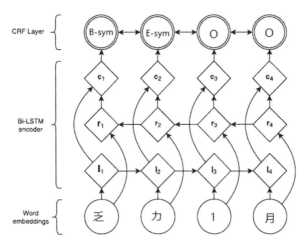

Fig. 1. BiLSTM-CRF structure

2.2 Lattice LSTM-CRF

Character-based model can avoid word segmentation errors, but it can also lose the semantic information contained in the word sequence in the context. Lattice model uses lattice LSTM to represent lexicon words in input sentences, so as to integrate potential word information into character based LSTM-CRF model [7], as shown in Fig. 2. Lattice structure can use character and word sequence information, gating structure to select the most relevant characters and words to get better recognition results [9].

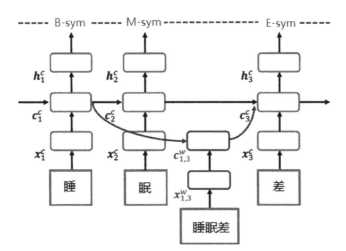

Fig. 2. Lattice LSTM-CRF model

2.3 BERT

The full name of the BERT model is bidirectional encoder representation from transformers, and the structure [8] is shown in Fig. 3. BERT uses two-way transformer structure, which can obtain information in two directions before and after sentences at the same time. It uses multi-layer self-attention mechanism to replace the traditional RNN, CNN neural network, and solves the thorny long-term dependence problem in natural language processing.

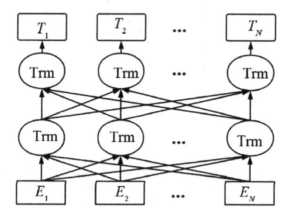

Fig. 3. BERT pre-trained language model

3 Experiment

3.1 Data

The data set used in this study came from the Hubei Provincial Hospital of Traditional Chinese Medicine. A total of 131 electronic medical records were used. The current medical history and main complaints in the admission records were used to label the data set, and the doctor marked the symptoms. Among them, 103 medical records were used for training set, including 1,025 symptom entities; 14 medical records were used for verification set, including 143 symptom entities; 14 medical records were used for test set, including 173 symptom entities.

3.2 Sequence Labeling

In this paper, BMEO is used to label, B represents the starting position of the entity, e represents the ending position of the entity, M represents the middle position of the entity, O represents the non-entity.

3.3 Evaluation Index

The precision (P), recall (R) and F1 value are used to evaluate the performance of the model, in which

$$P = \frac{the\ number\ of\ correctly\ identified\ entities}{the\ total\ number\ of\ identified\ entities} \times 100\%, \tag{1}$$

$$R = \frac{the\ number\ of\ correctly\ identified\ entities}{the\ total\ number\ of\ all\ entities} \times 100\%, \tag{2}$$

$$F_1 = \frac{2 \times P \times R}{P + R} \times 100\%. \tag{3}$$

3.4 Parameter Setting

Set the BiLSTM-CRF model initialization vector dimension to 100 dimensions, the learning rate to 0.001, batch-size to 32, and the number of iterations to 100. The Lattice LSTM-CRF model uses a pre-trained character vector set and word vector set giga-word_chn.all.a2b.uni.ite50.vec, which is a vector set trained by the Chinese corpus Giga-word using the Word2vec tool after a large-scale standard word segmentation, with 100 iterations, an initial learning rate is 0.015 and the decay rate is 0.05. The BERT model uses the Chinese version of BERT-Base, the batch size is set to 32 in the experiment, the learning rate is 2e−5, and the epochs is 10.

3.5 Result

The experimental results of each model are shown in Table 1.

Table 1. Experimental results of different models

Model	P (%)	R (%)	F1 (%)
BiLSTM-CRF	89.33	77.94	83.25
Lattice LSTM-CRF	92.16	61.44	73.73
BERT	89.94	88.27	89.10

As can be seen from the experimental results,

(1) As the mainstream named entity recognition model, the precision of BiLSTM-CRF is 89.33%, which is slightly lower than that of Bert model, the recall and F1 value are higher than lattice model, but the precision is lower than lattice model.
(2) The precision of Lattice LSRM-CRF is the highest, reaching 92.16%. The lattice model integrates word information into character features, which can better repre-sent word information. However, the recall is the lowest, which may be related to

the pre-trained word vectors used. The vector set based on the large-scale standard word segmentation Chinese corpus training may be more suitable for the general field, rather than the Chinese medicine field. It may be better to use large-scale TCM corpus to train word vectors.

(3) The recognition result based on the BERT model has the highest recall and F1 value, with a recall of 88.27% and F1 of 89.10%, indicating that the BERT word vector can better represent the semantic information of the word.

4 Conclusion

In this study, three models of BiLSTM-CRF, Lattice LSTM-CRF, and BERT, were used to identify the named entities of symptoms in the electronic medical records of traditional Chinese medicine. In general, the BERT model has the best experimental results, with an accuracy rate of 89.94%, a recall rate of 88.27%, and an F1 value of 89.10%. Recognition of named entities of electronic medical records of traditional Chinese medicine is still in the development stage. This experiment only identified symptomatic entities, and also included various entities such as disease, examination, inspection, and surgery in the electronic medical record. In the future, the scope of entity recognition will be expanded, and more corpora will be obtained for model training to improve the recognition effect, so as to structure the electronic medical record.

Acknowledgements. This study was supported by the Health Commission of Hubei Province Guiding Project (#WJ2019F185).

References

1. Li, B., Kang, X., et al.: Named entity recognition in Chinese electronic medical records using Transformer-CRF. Comput. Eng. Appl. **56**(5), 153–159 (2020)
2. Lample, G., Ballesteros, M., Subramanian, S., Kawakami, K., Dyer, C.: Neural architectures for named entity recognition. In: Proceedings of the 2016 Conference of the North American Chapter of the Association for Computational Linguistics: Human Language Technologies, pp. 260–270 (2016)
3. Sun, C., Xie, Q.: Discussion on methods of terminology recognition in TCM medical records. Chin. J. Libr. Inf. Sci. Tradit. Chin. Med. **44**(2), 1–5 (2020)
4. Li, M., Liu, Z., Yao, Y.: LSTM-CRF based symptom term recognition on traditional Chinese medical case. J. Comput. Appl. **38**(S2), 42–46 (2018)
5. Gao, S., Jin, P., Zhang, D.: Research on named entity recognition of TCM classics based on deep learning. Technol. Intell. Eng. **5**(01), 113–123 (2019)
6. Zhang, Y., Guan, B., et al.: Study on the entity extraction of traditional Chinese medicine on the basis of deep learning. J. Med. Inform. **40**(02), 58–63 (2019)
7. Zhang, Y., Yang, J.: Chinese NER using lattice LSTM. In: Proceedings of the 56th Annual Meeting of the Association for Computational Linguistics (Long Papers), pp. 1554–1564 (2018)

8. Devlin, J., Chang, M., Lee, K., et al.: BERT: pre-training of deep bidirectional transformers for language understanding. In: Proceedings of the Conference of the North American Chapter of the Association for Computational Linguistics: Human Language Technologies, pp. 4171–4186 (2019)
9. Pan, C., Wang, Q., et al.: Chinese electronic medical record named entity recognition based on sentence-level lattice-long short-term memory neural network. Acad. J. Second Mil. Med. Univ. **40**(05), 497–506 (2019)

AHIAP: An Agile Medical Named Entity Recognition and Relation Extraction Framework Based on Active Learning

Ming Sheng[1], Jing Dong[2(✉)], Yong Zhang[1], Yuelin Bu[3], Anqi Li[4], Weihang Lin[5], Xin Li[6], and Chunxiao Xing[1]

[1] BNRist, DCST, RIIT, Tsinghua University, Beijing 100084, China
{shengming,zhangyong05,xingcx}@tsinghua.edu.cn
[2] University of Queensland, Brisbane, QLD 4072, Australia
j.dong1@uq.net.au
[3] Beijing University of Posts and Telecommunications, Beijing 100876, China
graceyuelin.bu@gmail.com
[4] Beihang University, Beijing 100191, China
anqili99@hotmail.com
[5] Beijing Foreign Studies University, Beijing 100089, China
oscar.lam.bfsu@gmail.com
[6] Beijing Tsinghua Changgung Hospital, School of Clinical Medicine, Tsinghua University, Beijing, China
Horsebackdancing@sina.com

Abstract. Knowledge graph plays a significant role in many domains for providing a wide range of assistance. In the medical domain, clinical guidelines, academic papers, Electronic Medical Records (EMRs) and crawled data from the Internet contain essential information. However, those data are usually unstructured but vital to knowledge graph construction. The construction of knowledge graph using unstructured data requires a large number of medical experts to participate in annotations based on their prior experiences and knowledge. Knowledge graphs' quality highly depends on the performances of medical named entity recognition and relation extraction that are both based on data annotation. However, faced with handling such a large amount of enormous data, manual labelling turns out to be a high labor cost task. Besides, the data is generated rapidly, requiring us to annotate and extract quickly to keep the pace with the data accumulation. Therefore, we propose a named entity recognition and relation extraction framework, AHIAP, to solve these problems mentioned above. AHIAP uses active learning method to reduce the labor cost of the annotation process while maintaining the annotation quality. There are two modules in AHIAP, an active learning module for reducing labor cost and a measurement module to control the quality. By using active learning, AHIAP only takes 200 samples to get to the accuracy of 70%, whereas the standard learning strategy takes 4000 records to get the same accuracy.

Keywords: Health knowledge graph · Doctor-in-the-loop · Platform · Healthcare · Health informatics

© Springer Nature Switzerland AG 2020
Z. Huang et al. (Eds.): HIS 2020, LNCS 12435, pp. 68–75, 2020.
https://doi.org/10.1007/978-3-030-61951-0_7

1   Introduction

Knowledge graphs collect a massive amount of interrelated facts that connect different concepts and instances, and can be transformed into practical knowledge [1]. These linked data triples can be queried by users [2], and support doctors to make diagnostic decisions [3] or develop medical applications while supporting patients to find symptoms-relevant information [4]. Researchers have paid a great amount of effort into the realms of constructing knowledge graphs. There are several developed knowledge graphs available in medical domain like Linked Life Data (LLD), which connecting over 20 bio-databases. However, those health knowledge graphs focus only on storing concept information.

In these health knowledge graphs, RDF triples are stored to represent the knowledge, and there are three types of information stored as nodes: entity, event and concept. We define the knowledge graph with concept nodes as Concept Knowledge Graph (CKG). Correspondingly, we define the knowledge graph with entity nodes and event nodes as Instance Knowledge Graph (IKG). Based on that, we describe the knowledge graph, including both CKG and IKG as Factual Knowledge Graph (FKG) [5]. In the medical domain, IKG contains instance data such as the records of medical, which are useful for further analysis. But these data sources are generally unstructured, in which the knowledge needs to be extracted manually with dramatic labor cost.

To reduce the labor cost, automatic named entity recognition and relation extraction are adopted. Maya Rotmensch et al. [6] presented a methodology for constructing a health knowledge graph using automatic entity extraction from unstructured data. But the mechanical method to do so still requires preprocessed data and a lot of time in model training [7]. Besides, without the doctors to provide useful prior knowledge and measure the process, the quality of the automatic result is relatively unreliable [8].

To solve the labor cost problem, we propose a medical named entity annotation and relation extraction framework AHIAP, which implements active learning to reduce human participation workloads during the medical unstructured data annotation process, and is further combined with "doctor-in-the-loop" methodology [9] to maintain the quality of entity annotation and relation extraction result.

This paper is organized as follows. In Sect. 2 we introduce the related work in the relevant field. In Sect. 3 we present the framework and workflow of AHIAP. In Sect. 4 we show the details of the modules used in this framework, introduce how they reduce the labor cost and perform the quality control. In the end, we summarize the paper and propose future work in Sect. 5.

2   Related Work

In Table 1, nine frameworks that can be used in named entity recognition and relation extraction tasks are listed. They are compared on human participation method and labor cost level. As shown from the table, only three frameworks combine both machine and human effort to accelerate the annotation process with a reliable result. Among all the frameworks listed, only the WebAnno provides full auto annotation but it is only available for project manager and administrators. Most of the named entity recognition and relation extraction frameworks are purely manual.

Table 1. Named entity recognition and relation extraction frameworks

Framework name	Human participation method	Labor cost level
Doccano [10]	Manual	High
BRAT [11]	Manual	High
Prodigy [12]	Semi-auto	Middle
YEDDA [13]	Manual	High
DeepDive:Mindtagger [14]	Manual	High
Anafora [15]	Manual	High
WebAnno [16]	Auto/manual	Low/High
MAE [17]	Manual	High
INCEpTION [18]	Semi-auto	Middle

3   The Framework of AHIAP

The Framework of AHIAP

As Fig. 1 shows, the framework contains three parts: (1) The data source of AHIAP provides unstructured medical data to be annotated; A high-quality health CKG is also an input to provide annotation labels. (2) The building modules. (3) The output of this framework is the high quality annotated medical unstructured data and can be further used to construct high-quality IKG.

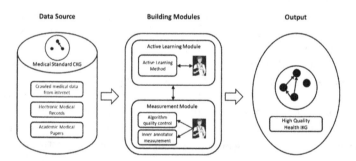

Fig. 1. The framework of AHIAP.

The Workflow of AHIAP

As shown in Fig. 2, in the workflow of AHIAP, the medical unstructured data is taken as input into the active learning module, doctors who are assigned as annotators are asked to annotate a small set of data randomly selected from input dataset. Then, those data are sent to the active learning algorithm to train the model. After initializing a learning model, the algorithm periodically returns the unconfident auto annotation result to the annotator, asks them to prove.

With the model keeping convergence, it becomes more and more accurate and requires less human effort in correction. The measurement module is supervised during the entire process. The fine trained model can automatically extract medical unstructured data and generate high-quality health IKG with almost zero labor cost.

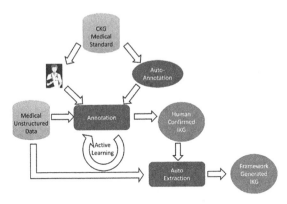

Fig. 2. The detailed workflow of AHIAP to annotate medical unstructured.

4 Modules in AHIAP

In this section, we describe the active learning process of the active learning module in detail, and explain how the mechanism of the measurement module works.

4.1 Deep Active Learning Module

Algorithms that involve humans' interventions can be defined as "human-in-the-loop". Human-in-the-loop has been applied to many aspects of artificial intelligence like named entity recognition [19] and rules learning [20] to improve the performance. Active learning is a machine learning method that involves the human-in-the-loop methodology.

In AHIAP, we use Shen's work [21] to implement an active learning model. When the active learning model is compared with other algorithms, pure deep learning needs a larger labelled dataset to perform well, but when it comes to small datasets, the advantage is less obvious. Meanwhile, expecting better performance with less manually labelling work, active learning methods seek to select a subset of examples that can critically improve the model before ask the annotators to label them.

The deep learning method in our experiment is a CNN-CNN-LSTM architecture including character-level encoder, word-level encoder and tag decoder. The input unstructured data with the low rank will be chosen for active learning use sequence tagging.

As we can see from Fig. 3, using active learning it only takes 200 samples to get to the accuracy over 70%, whereas the standard learning takes more than 4000 records to get the same accuracy. As the number of samples increases, the performance of the model still remains stable. Besides this experiment also shows in the medical field active learning can reduce annotation cost and result in better quality predictors in same time.

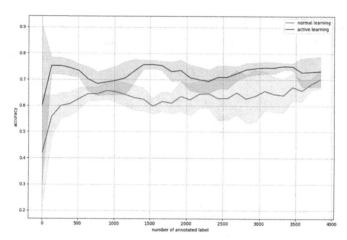

Fig. 3. The converged result between using active learning and standard learning.

4.2 Active Learning Based Named Entity Recognition and Relation Extraction Process

During the learning process, active learning algorithm iteratively queries the most informative instances to manual verification and revision. The appropriate selection of instances in each epoch ensures the cost of manual work being limited in a relatively low level.

Start-up Procedure for Active Learning Process
At the start-up of an annotation assignment, annotation manager initializes it, and determines the field of this assignment; chooses the target medical documents that need to be annotated; a CKG is used for medical standard, and assigned to the doctors. Then, the framework pushes part of randomly selected medical document to doctor, and let them label the data. The labeled data is sent to the measurement module before being transferred to a "storage of training data". The initially trained model is generated through the startup procedure.

Loop Procedure for Active Learning Process
After the startup, there is a loop of the active learning. In this process, the trained machine learning model tries to automatically perform NER on those unstructured medical data not in the training storage, resulting in a machine labelled records. Next, those records are applied for uncertainty-based sampling strategy and calculate the confidence to every machine labelled data. A certain amount of data with the lowest confidence is passed on to the doctors for the annotation. The doctor can choose to accept those annotation results labelled by model or re-annotate them again. Doctors' annotation results are transmitted to the training storage if they can pass the measurement tool. The machine learning module updates based on the new training storage. Finally, the framework starts the next cycle of a loop by applying the trained model to the unstructured medical data

out of the storage. The start-up and loop procedure are where active learning take place and reduce the labor cost.

Termination Procedure for Active Learning Process
The loop terminates once the annotation manager regard that the performance of the model is good enough. The data in training storage and the rest of the machine labelled data is moved to the result storage and becomes the final result of this annotation assignment.

With the growing of dataset to annotate, only a few of data need to be manually processed. Therefore, the labor cost is reduced using this module.

4.3 Measurement Module

In the medical field, system stability is sensitive because they may cause not only some ethical problems but also bring severe medical accidents to patients. Therefore, we must decrease the mistakes of the framework.

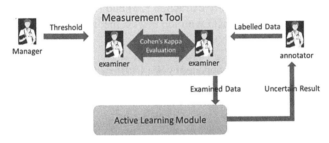

Fig. 4. Workflow of the inner annotator agreement measurement system.

Inner Annotator Agreement Measurement System
To measure the quality of annotated data, the mistakes from the annotator should be minimized. Therefore, we implement an inner annotator agreement measurement system shown as Fig. 4, Cohen's Kappa, has been proved as a very effective agreement measurement evaluation method [22]. We apply that evaluation between the examiners who measure the labelled result from annotator, and only send the examined data which pass the evaluation score threshold to the active learning module. Need to mention that in the framework of AHIAP, examiners, annotation managers and annotators should be doctors with prior knowledge due to the particularity of the medical field.

Up-to-date CKG
This framework using reliable CKG to provide medical standards. During the annotation process, the doctors are asked to choose from CKG standard to annotate on the target corpus rather than self-defined one, which helps doctors to annotate the medical unstructured data and produce reliable and standardized annotation results. Modification to the standards will be strictly restricted.

By applying direct mapping between annotation result and model prediction result with the up-to-date CKG, the high-quality FKG will be generated and ready to use for providing further help in the medical field.

5 Conclusion

In this paper, we propose AHIAP, an agile medical named entity recognition and relations extraction framework used for constructing high-quality health knowledge graph with low labor cost. In AHIAP, we develop two modules to make this framework to require less labor and keep accurate at the same time. The active learning module involves machine learning method with human-in-the-loop mechanism. It makes the trained machine learning model to converge with less data, and eventually, reduces the labor cost in the annotation process. The measurement module ensures the quality of annotation work in real-time by supervising any modification to the data and performing quality control using inner annotator measurement method.

In the future, we will apply AHIAP in the construction of knowledge graphs in more fields for establishing efficient medical support applications, such as cost prediction [23], document analysis [24], entity extraction [25] and recommendation [26]. We are also planning to build a larger CKG with the newest medical standards to improve the performance of this framework.

Acknowledgement. This work was supported by NSFC (91646202), National Key R&D Program of China (2018YFB1404401, 2018YFB1402701).

References

1. Pujara, J., Miao, H., Getoor, L., Cohen, W.: Knowledge graph identification. In: Alani, H., et al. (eds.) ISWC 2013. LNCS, vol. 8218, pp. 542–557. Springer, Heidelberg (2013). https://doi.org/10.1007/978-3-642-41335-3_34
2. Verborgh, R., et al.: Triple Pattern Fragments: a low-cost knowledge graph interface for the Web. J. Web Semant. **37**, 184–206 (2016)
3. Donnelly, K.: SNOMED-CT: the advanced terminology and coding system for eHealth. Stud. Health Technol. Inform. **121**, 279 (2006)
4. Agarwala, R., et al.: Database resources of the national center for biotechnology information. Nucleic Acids Res. **45**, D12–D17 (2017)
5. Sheng, M., et al.: DEKGB: an extensible framework for health knowledge graph. In: ICSH, pp. 27–38 (2019)
6. Rotmensch, M., Halpern, Y., Tlimat, A., Horng, S., Sontag, D.: Learning a health knowledge graph from electronic medical records. Sci. Rep. **7**, 1–11 (2017)
7. Lample, G., Ballesteros, M., Subramanian, S., Kawakami, K., Dyer, C.: Neural architectures for named entity recognition. arXiv preprint arXiv:1603.01360 (2016)
8. Giorgi, J.M., Bader, G.D., Wren, J.: Towards reliable named entity recognition in the biomedical domain. Bioinformatics **36**, 280–286 (2020)
9. Sheng, M., et al.: DocKG: a knowledge graph framework for health with doctor-in-the-loop. In: Wang, H., Siuly, S., Zhou, R., Martin-Sanchez, F., Zhang, Y., Huang, Z. (eds.) HIS 2019. LNCS, vol. 11837, pp. 3–14. Springer, Cham (2019). https://doi.org/10.1007/978-3-030-329 62-4_1

10. doccano - Document Annotation Tool. https://doccano.herokuapp.com/. Accessed 11 June 2020
11. brat rapid annotation tool. https://brat.nlplab.org/
12. Prodigy · An annotation tool for AI. Machine Learning & NLP. https://prodi.gy/
13. Jie, Y., Yue Z., Linwei L., Xingxuan L.: YEDDA: a lightweight collaborative text span annotation tool. In: ACL 2018, pp. 31–36 (2018)
14. Deepdive. https://github.com/HazyResearch/deepdive. Accessed 11 June 2020
15. Chen, W., Styler, W.: Anafora: a web-based general purpose annotation tool. In: NAACL, pp. 14–19 (2013)
16. Eckart de Castilho, R., et al.: A web-based tool for the integrated annotation of semantic and syntactic structures. In: LT4DH Workshop, pp. 76–84 (2016)
17. Multi-document Annotation Environment. http://keighrim.github.io/mae-annotation/
18. Klie, J.-C., Bugert, M., Boullosa, B., Eckart de Castilho, R., Gurevych, I.: The INCEpTION platform: machine-assisted and knowledge-oriented interactive annotation. In: ACL, pp. 5–9 (2018)
19. Coelho da Silva, T.L., Magalhães, R.P., et al.: Improving named entity recognition using deep learning with human in the loop. In: EDBT, 594–597 (2019)
20. Yang, Y., Kandogan, E., Li, Y., Sen, P., Lasecki, W.S.: A study on interaction in human-in-the-loop machine learning for text analytics. In: CEUR Workshop (2019)
21. Shen, Y., Yun, H., Lipton, Z.C., Kronrod, Y., Anandkumar, A.: Deep active learning for named entity recognition. arXiv preprint arXiv:1707.05928 (2017)
22. Vieira, S.M., Kaymak, U., Sousa, J.M.C.: Cohen's kappa coefficient as a performance measure for feature selection. In: WCCI 2010. pp. 1–8. IEEE (2010)
23. Zhao, K., et al.: Modeling patient visit using electronic medical records for cost profile estimation. In: DASFAA, pp. 20–36 (2018)
24. Tian, B., Zhang, Y., Wang, J., Xing, C.: Hierarchical inter-attention network for document classification with multi-task learning. In: IJCAI, pp. 3569–3575 (2019)
25. Wang, J., Lin, C., Li, M., Zaniolo, C.: Boosting approximate dictionary-based entity extraction with synonyms. Inf. Sci. **530**, 1–21 (2020)
26. Zhao, K., et al.: Discovering subsequence patterns for next POI recommendation. In: IJCAI, pp. 3216–3222 (2020)

Medical Information System

An Integrated Knowledge Graph for Microbe-Disease Associations

Chengcheng Fu[1,2], Ran Zhong[1,2], Xiaobin Jiang[1,2], Tingting He[1,2], and Xingpeng Jiang[1,2(✉)]

[1] Hubei Provincial Key Laboratory of Artificial Intelligence and Smart Learning, Central China Normal University, Wuhan 430079, Hubei, PR China
xpjiang@mail.ccnu.edu.cn
[2] School of Computer, Central China Normal University, Wuhan 430079, Hubei, PR China

Abstract. Following the rapid advances of the human microbiome, the importance of micro-organisms especially bacteria is gradually recognized. The interactions among bacteria and their host are particulary important for understanding the mechanism of microbe-relate diseases. This article mainly introduces an explorative study to extract the relations between bacteria and diseases based on biomedical text mining. We have constructed a Microbe-Disease Knowledge Graph (MDKG) through integrating multi-source heterogeneous data from Wikipedia text and other related databases. Specifically, we introduce the word embedding obtained from biomedical literature into traditional method. Results show that the pre-trained relation vectors can better represent the real associations between entities. Therefore, the construction of MDKG can also provide a new way to predict and analyse the associations between microbes and diseases based on text mining.

Keywords: Microbes and diseases associations · Text mining · Knowledge graph

1 Introduction

Microorganisms live together with human, most of which are scattered throughout the oral cavity, skin, gastrointestinal tract and other parts of the body. As Human Microbiome Project (HMP) reported [1], the microbial community and metabolic pathway are different in each part of the healthy body. Microbes are also closely related to diabetes, premature delivery and inflammatory bowel disease [2]. Through genetic sequencing, construction of metabolic pathway and other interdisciplinary efforts, the dynamic status between microbes and hosts in healthy body are further depicted.

Microbiota is susceptible to environmental factors, the balance status of microbiota is easily disrupted by seasonal changes, viral invasion or smoking, so that human might suffer immune arthritis, pneumonia or other diseases. Besides the autoimmune diseases, the effects of mental disorders have become increas-

© Springer Nature Switzerland AG 2020
Z. Huang et al. (Eds.): HIS 2020, LNCS 12435, pp. 79–90, 2020.
https://doi.org/10.1007/978-3-030-61951-0_8

ingly important in recent years. Human with mental and physical pressure might get mental diseases, such as depression and anxiety. The channel between gut microbiota and the central nervous system, as well as the brain-gut-microbe, can also affects social behavior and status of human [3]. This can explain the causes of several mental diseases, like depression or autism, and accelerate the research work of new treatments. The accumulated evidences present a microbe-disease network linking complex diseases and microbes.

The massive biomedical literature provide the necessary conditions for text mining [4] to construct the microbe-disease network. The microbe-disease entity associations can be extracted automatically through machine learning and deep learning models. Then, these association triples could be integrated as a knowledge graph, which is a convenient way to store and retrieve the non-relational data. By learning the representation of graph we can predict more potential nodes and relationships and gives decision support for clinical consultation, drug design [5] and health or nutrition management. In this paper, we introduces an explorative study to extract the relations between bacteria and diseases from text mining Wikipedia text and other related databases. Furthermore, a Microbe-Disease Knowledge Graph (MDKG) is constructed for novel association prediction.

The following content is divided into five parts: Sect. 2 introduces the several related researches. Section 3 describes the work flow, which includes text mining, data integration and representation learning of MDKG. Section 4 presents the applications of MDKG. Finally, Sect. 5 makes a summary and discusses further researches.

2   Related Work

This section will compare several relevant researches about the databases of microorganisms and diseases, introduce some specific domain knowledge graph and discuss their characteristics.

2.1   Other Databases

In terms of the microbial study, many researchers publish the databases. In the meanwhile, there are many evaluation tasks for entity relation extraction. Such as BioCreative, BioNLP, SemEval, etc. These competitions have greatly promoted the progress of the research about biomedical literature mining.

In 2016, Wei Ma et al. [6,7] read nearly 100 biomedical literatures manually. They released the MicroPattern contains 47 kinds of microorganisms and 37 kinds of diseases. In 2018, Janssens et al. [8] linked external data to standardize microorganisms and disease entities. And they integrated biomedical literature mining with electronic medical record. There are also many sub-tasks such as named entity recognition and relationship extraction in microbes and disease [9,10].

However, extracting the relationship artificially couldn't be updated in large-scale. And there has hardly database about bacterial attributes knowledge. From the above, this paper proposes a research on the construction of an integrated knowledge graph about the microbes and diseases associations.

2.2 Knowledge Graph

Wikipedia integrates knowledge in various fields as a public resource platform. In recent years, some researchers have begun to use Wikipedia for biomedical research. Zinovyev A et al. [11] used the hidden link in Wikipedia to study certain associations between proteins. Rollin G et al. [12] studied the outbreak logs of infectious diseases in various countries through Wikipedia and put forward reliable prediction.

With the expansion of data volume and information retrieval dimension, knowledge graph has become a tool for storage and fusion of knowledge [13]. Malas et al. [14] utilized media-related terms, as well as context-semantic words, for relevant drug design in terms of specific disease. In the domain of traditional Chinese medicine, Tong Yu [15] applied ontology learning and knowledge fusion. They also constructed a knowledge graph for Chinese medicine treatment. These methods constructed their domain knowledge graph but overlooked contextual relationships, such as semantic or syntactic associations. Therefore, this paper proposes a method for construction and reasoning of microbes and diseases knowledge graph.

3 Construction of MDKG

This section discuss the details of the study. The work flow covers three steps. First of all, we adopt a text mining tool extracting relationships. Secondly, In this paper, there are numerous multi-source heterogeneous data that is integrated. Finally, we utilize our knowledge graph for link prediction.

3.1 Framework of MDKG

MDKG mainly contains three parts of content, as shown in Fig. 1. The data preparation contains the text mining on Wikipedia and data integration, which is the foundation of a domain knowledge about diseases and microbes. In text mining, the input is the sentences from pages of Wikipedia. Through a series of natural language processing, the output is triples about diseases and microbes associations. In the second phase, data integration contains other sources (such as open data from competition, specific databases, etc.). And the output is structured tables of database. Finally, we utilize these data to design the MDKG, and apply our knowledge graph into several applications. In the next subsections all the three steps are discussed in detail.

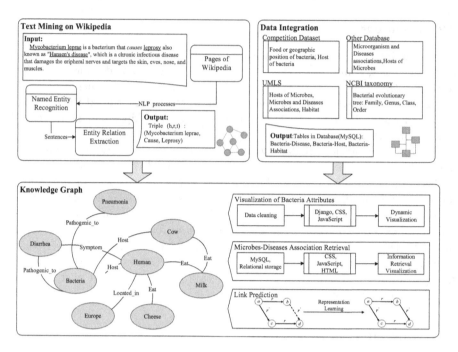

Fig. 1. Work flow of Microbes-Diseases knowledge graph construction. Text Mining on Wikipedia: the triple of microbes and diseases associations from Wikipedia text mining. Data Integration: the integration of multi-source heterogeneous database. The output of both two steps are the input of the final construction. Knowledge Graph: the construction and application of microbes and diseases knowledge graph.

3.2   Text Mining

Wikipedia consist of thousands of topics such as society, life, people, history, etc. This paper split text of Wikipedia pages into sentences through natural language processing processes. Then, we perform named entity recognition and relationship extraction on the sentences and obtains the interaction triplets.

Preprocessing. Wikipedia has compressed all the body pages as the dump file. The dump file contains entries, templates, picture descriptions and basic meta-pages. First of all, we need to preprocess through the following steps:

1. Remove pictures and tables in the dump file.
2. Use SPARQL query in Wikidata obtaining the contextual relationship and synonyms between bacteria and disease. Wikidata is a structured knowledge base linked Wikipedia, which involves knowledge in the form of triples.
3. Filter out the bacterial text through Wikidata and convert to BioC format.

After the above processes, we get 436,196 entries for bacteria, 8,483 disease entries and 6,000 bacterial pages.

Entity Recognition. We apply a tool named Kindred for named entity recognition. Kindred is a toolkit developed by Stanford, which provides interfaces for biomedical text mining. The input of Kindred are clinical nicknames of diseases from Wikidata and abbreviated names of bacteria from Unified Medical Language System (UMLS) as well as their standard names. The output is the sentences which contain at least one bacterial name and disease name. After identifying entities effectively, we acquire 2089 target sentences.

Relation Extraction. Each target sentence identifies the Wikidata number of the bacteria. We filter the target sentences by keywords 'infect', 'cause', 'pathogen' that as many sentences as possible containing information on bacterial pathogenicity are selected. Then, we apply the Kindred relation classifier for relation extraction. The input is the target sentences, and the output is the triples of microbes and diseases associations.

We obtains 132 kinds of bacteria, 152 kinds of diseases and 208 pairs of bacteria and diseases associations from Wikipedia. In this paper, we only discuss the species-level interaction. Compared with other databases such as MicroPattern [6] and National Center for Biotechnology Information (NCBI), our method get effective increment.

3.3   Data Fusion

We have integrated NCBI and UMLS, related competitions and other databases. The data sets contain more than 9000 microorganisms and 23 types of bacterial attributes. After removing some items with many missing values and redundant ID features, we get 7765 species and 11 attribute types.

In this paper, clustering and association analysis methods are used to comprehending bacterial features. As shown in Fig. 2. (a) We selects five attributes of bacteria: gram strain, salinity of bacteria, oxygen requirement of bacteria, temperature range of growth environment, and habitat of bacteria, which is for cluster analysis of the pathogenicity of bacteria. (b) We selected GC content, salinity of bacteria, oxygen requirement, gram strain, habitat and pathogenicity of bacteria to draw a heat map of the relationship between each attribute. The number and color about each box indicate the strength of the correlation between the two attributes. (c) We turn the diseases caused by bacteria in different hosts into visualization.

After these analysis, we can draw the conclusion. On the one hand, the same bacteria can cause different diseases in an organism. For example, Chlamydia trachomatis can cause gonorrhea and blindness in the human. On the other hand, the same microbe can also infect different hosts. Such as Pasteurella multocida can cause atrophic rhinitis in human and pasturellosis in sheep. Therefore, this paper further recognizes that bacterial attributes can affect each other and influence the balance of the body.

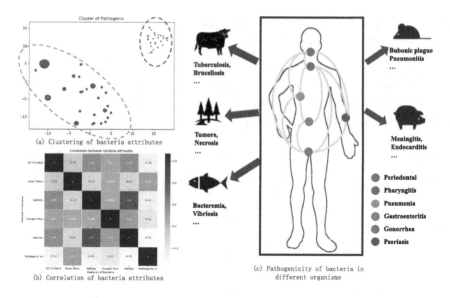

Fig. 2. Bacterial attributes analysis and fusion. (a) Using bacterial attributes to cluster and get its association with bacterial pathogenicity; (b) Correlation between the six attributes of bacteria; (c) Pathogenicity of bacteria in different parts of the human, cattle, mice, fish, pig and plants.

3.4 Knowledge Inference

Table 1. Entity associations type and number.

The head	The tail	Association type	Counts
Species	Diseases	Pathogenic to	2594
Species	Bacterial host (animal, plant or microorganism)	Host in	1743
Species	Bacterial symbiotic community	Group in	7586
Species	Bacterial habitat (food or geographical conditions)	Locate in	1279
Species	Bacterial genus	Genus is	2196
Bacterial genus	Bacterial family	Family is	2171
Bacterial family	Bacterial order	Order is	2171
acterial order	Bacterial class	Class is	2166

For further enrichment of the associations in the knowledge graph, this paper utilize representation learning method for knowledge inference. Firstly, we apply several traditional methods in MDKG for comparison, and the method with better effect has been selected through adjusting related parameters. Secondly, we introduce some improvement strategies in this model. Table 1 shows 8 relationship types based on bacterial association reasoning, and the distribution of each relationship type is relatively uniform. The data set contain 21,905 edges of associations, 9,832 entity types and 7 relationships. We divide the data set under 8:1:1. As shown in Table 2.

Table 2. Division of datasets.

	The head	The tail	Association type	Associations	Proportion
Train set	7865	7749	8	17524	8
Test set	991	976	8	2191	1
Validation set	983	965	8	2191	1

First of all, we define the triple (head, relation, tail) as (h, r, t), which satisfies $h, t \in E$ and $r \in R$, as input to the model. Each method will learn the representation of these triples. Traditional methods are mainly divided into three categories.

1. The bilinear model. In 2011, RESCAL [16] introduced tensor decomposition to represent multiple binary relationships. They took a full-rank matrix M_r as relationship and a word vector A as entity. $X_r \approx AM_rA^T$ represents a triple. And the mapping function is:

$$f_r(h, r, t) = hM_rt^T \tag{1}$$

Most of the bilinear models are general. DistMult is a multiplicative model [17], which was proposed in 2015. It represents triple as a low-dimensional embedding. ComplEx [18] applies imaginary part embedding for learning the asymmetric and inverse relationships. There also has ANALOGY framework [19], which added a latent semantic embedding.

2. The neural network. We select the most common graph convolutional neural network (GCN) [20] as a representation in deep learning. Similar to the above methods, GCN expresses the relationship as a diagonal matrix R_r, and the scoring function f_r can be expressed as:

$$f_r(h, r, t) = e_h^T R_r e_t \tag{2}$$

This method can learn the characteristics of neighbor nodes through a normalized and cumulative learning, which can continuously update the representation of the node. Where e_r represents the hidden state of a node in a layer l of the neural network, so that can be defined as:

$$e_i^{(l+1)} = ReLU\left(\sum_{r \in R}\sum_{j \in N_i^r}\frac{1}{c}W_r^{(l)}e_j^{(l)} + W_0^{(l)}e_i^{(l)}\right) \tag{3}$$

Where N_i^r represents the neighbor node of the node i under the relationship r. Here, c represents the constant for normalization, which can be learned or defined in advance.

3. The translation model. In 2013, Bordes et al. [21] proposed the TransE, which translates relationship vectors into a head to tail entity mapping approximately. And the mapping function f_r can be defined as:

$$f_r(h, t) = -\|h + r - t\|_{1/2} \qquad (4)$$

After experimental verification, as shown in Table 3, TransE performs better on the MDKG. Inspired by word embedding in natural language processing, this paper introduces the representation of text-based relationship vectors into TransE, as shown in Fig. 3.

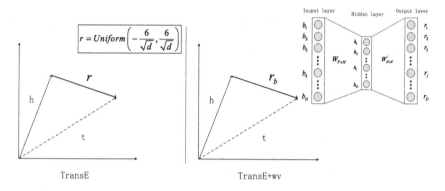

Fig. 3. Comparison structure of TransE and TranE+wv. TransE applies the random initialization to define the relation vectors r; TransE+wv applies the word2vec to represent the pre-trained relation vectors.

Table 3. The training results of models.

Models	MDKG				
	MRR (%)		Hits@N (%)		
	Raw	Filtered	1	3	10
GCN	32.2	41.2	39.9	41.4	44
ComplEx	37.4	44.5	40.3	43.3	57.8
Distmult	38	45.3	40.4	47.3	58.3
ANALOGY	42.8	49.4	44.6	51.1	58.9
RESCAL	38.1	44.7	40.3	45	58.1
TransE	43.5	49.9	45.5	51.7	59.6
textbfTransE+wv	**45.5**	**53.1**	**47.2**	**56.4**	**63.4**

Firstly, we retrieves and downloads 10,000 abstracts from PubMed with keywords 'bacteria', 'disease' and 'human'. Secondly, through removing punctuation and stop words and other preprocessing steps, we use Word2Vec [22] to train these sentences. Then, we convert the random initialization representation r into a relationship vector r_b pre-trained from biomedical text. Finally, we use the parameters $lr = 0.08, d = 100, gamma = 1.0$ and get the best performance of TransE+wv on the data set. The results of each model are shown in Table 3.

We utilize mean reciprocal rank (MRR) [23] and Hits at N(Hits@N) as evaluation indicators. MRR can be represent as:

$$MRR = \frac{1}{|W|} \sum_{i=1}^{|W|} \frac{1}{rank_i} \tag{5}$$

Where $rank_i$ represents the rank of the correct answer for candidate i triplet. The results show that TransE+wv performs best on the data set. The bilinear model might ignore the implicit contextual semantic relationship, and the neural network model might be more suitable for large-scaled data. That might explain why they are poor on our data set. TransE with pre-trained relation embedding can represent associations better and obtain best results of link prediction.

4 Application

Our design about the applications of MDKG can be divided into three aspects, which could provide more convenience for searching, analysing and performing further experimental verification.

4.1 Visualization

In above works, the associations between attributes and pathogenicity of bacteria have been verified. This paper develops a dynamic analysis and display page of bacterial attributes and uses Echarts and AJAX technology for visualization.

The visualization interface exhibits 8 kinds of bacterial attributes: sequence, genome size, GC content, shape, microbiota size, bacterial salinity, oxygen requirement, temperature range. This page uses dynamic refresh technology drawing bar, pie and other charts, which could exhibit the distribution and characteristics intuitively.

4.2 Information Retrieval

The MDKG utilizes two different methods regarding to information retrieval. One is the query module, which displays all relevant information about diseases and microbes directly. The other method is the dynamic visualization retrieve module. As shown in Fig. 4, Green circles represent microbes, purple circles as diseases and grey ones represent other entities. The links between them represent their associations.

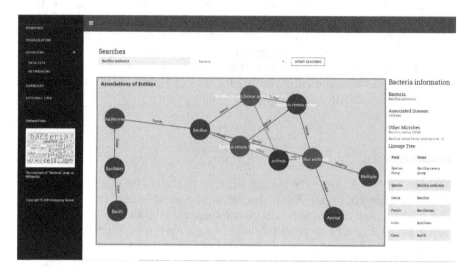

Fig. 4. The platform interface of knowledge graph. The right box is navigation bar of the platform. The middle box shows the associations about the entities. The right box exhibits the detail information about microbes. The platform can be accessed on https://github.com/ccszbd/MDKG.

Table 4. Examples of link prediction results based on TransE.

The head	Association type	Prediction sequence of the tail
Actinomyces odontolyticus F0309	Host in	**Human**, Cattle, Animal, Fish, Rodent, Horses, Thyroid carcinoma, Micrococcus, Porcine, Lactobacillus
Bilophila wadsworthia	Pathogenic to	Psoriasis, Crohn's Disease, Behcet's Disease, Pneumonia, **Cirrhosis**, Tuberculosis, Autism, Aggressive periodontitis, Irritable Bowel Syndrome, Lyme Disease
Escherichia coli O157:H7	Locate in	Micec, heese from dairy B, patients, Euprymna scolopes, Borrelia burgdorferi CA-11.2A, **Lettuce**, Snowella, Campylobacter hyointestinalis subsp. hyointestinalis LMG 9260, chicken nugget processing plant, susceptible individuals
Paenibacillus larvae subsp. larvae B-3650	Group in	Alphaproteobacteria, Actinobacteria, Gammaproteobacteria, **Firmicutes**, Other Bacteria, Betaproteobacteria, Chloroflexi, Cyanobacteria, S. equorum Mu2, Thermotogae

4.3   Link Prediction

The TransE+wv is employed to the link prediction in the knowledge graph. We apply the head entity and relationship for predicting the tail one. Table 4 lists the prediction results under 5 different relationship types. The prediction sequence show the potential optimal results of this triple, which correct entity is bolded.As it shows that even if the correct entity is not always in the first position, the prediction always satisfies common sense. For example, about the Bilophila wadsworthia's pathogenic link prediction. Near the correct answer 'Cirrhosis' are similar diseases, which indicates that they are very close in vector space. The knowledge graph of microbes and diseases constructed in this paper has various applications such as guiding the design of microbial experiments.

5   Conclusion

This paper carry out researches on the construction of knowledge graph based on diseases and microbes. However it still has directions for improvement: We can develop a bacterial attribute mining tool for expending the available database. In the link prediction, we can combine the semantic features and word features of the text. In conclusion, there are many research directions based on the knowledge graph research of bacteria and diseases. It is still not enough to comprehend the role of microorganisms in human. The interactions between bacteria itself, bacteria and antibiotics or viruses are also essential for research. The effect of bacteria on human health will always be the goals and directions of our researches.

Acknowledgement. This research is supported by National Key Research and Development Program of China (2017YFC0909502) and the National Natural Science Foundation of China (61532008 and 61872157). We also thanks to the support of Fundamental Research Funds for Central Universities (CCNU19ZN009).

References

1. Turnbaugh, P.J., Ley, R.E., Hamady, M., Fraserliggett, C.M., Knight, R., Gordon, J.I.: The human microbiome project. Nature **449**(7164), 804–810 (2007)
2. Zhao, L., et al.: Gut bacteria selectively promoted by dietary fibers alleviate type 2 diabetes. Science **359**(6380), 1151–1156 (2018)
3. Grenham, S., Clarke, G., Cryan, J.F., Dinan, T.G.: Brain-gut-microbe communication in health and disease. Front. Physiol. **2**, 94–94 (2011)
4. Lever, J., Zhao, E.Y., Grewal, J.K., Jones, M.R., Jones, S.J.M.: CancerMine: a literature-mined resource for drivers, oncogenes and tumor suppressors in cancer. Nat. Methods **16**(6), 505–507 (2019)
5. Lu, Y., Guo, Y., Korhonen, A.: Link prediction in drug-target interactions network using similarity indices. BMC Bioinformatics **18**(1), 39–39 (2017)

6. Ma, W., Huang, C., Zhou, Y., Li, J., Cui, Q.: MicroPattern: a web-based tool for microbe set enrichment analysis and disease similarity calculation based on a list of microbes. Sci. Rep. **7**(1), 40200–40200 (2017)
7. Ma, W., et al.: An analysis of human microbe-disease associations. Briefings Bioinformatics **18**(1), 85–97 (2017)
8. Janssens, Y., et al.: Disbiome database: linking the microbiome to disease. BMC Microbiol. **18**(1), 50–50 (2018)
9. Li, X., Fu, C., Zhong, R., Zhong, D., He, T., Jiang, X.: A hybrid deep learning framework for bacterial named entity recognition with domain features. BMC Bioinformatics **20**(16), 1–9 (2019)
10. Badal, V.D., et al.: Challenges in the construction of knowledge bases for human microbiome-disease associations. Microbiome **7**(1), 129 (2019)
11. Zinovyev, A., Czerwinska, U., Cantini, L., Barillot, E., Frahm, K.M., Shepelyansky, D.L.: Collective intelligence defines biological functions in Wikipedia as communities in the hidden protein connection network. bioRxiv, p. 618447 (2019)
12. Rollin, G., Lages, J., Shepelyansky, D.L.: World influence of infectious diseases from Wikipedia network analysis. IEEE Access **7**, 26073–26087 (2019)
13. Kwon, S., Yoon, S.: End-to-end representation learning for chemical-chemical interaction prediction. IEEE/ACM Trans. Comput. Biol. Bioinf. **16**(5), 1436–1447 (2019)
14. Malas, T.B., et al.: Drug prioritization using the semantic properties of a knowledge graph. Sci. Rep. **9**(1), 6281 (2019)
15. Yu, T., et al.: Knowledge graph for TCM health preservation: design, construction, and applications. Artif. Intell. Med. **77**, 48–52 (2017)
16. Nickel, M., Tresp, V., Kriegel, H.: A three-way model for collective learning on multi-relational data. In: International Conference on Machine Learning, pp. 809–816 (2011)
17. Yang, B., Yih, W., He, X., Gao, J., Deng, L.: Embedding entities and relations for learning and inference in knowledge bases. arXiv : Computation and Language (2014)
18. Hayashi, K., Shimbo, M.: On the equivalence of holographic and complex embeddings for link prediction. arXiv : Learning (2017)
19. Liu, H., Wu, Y., Yang, Y.: Analogical inference for multi-relational embeddings. arXiv : Learning (2017)
20. Schlichtkrull, M.S., Kipf, T., Bloem, P., Den Berg, R.V., Titov, I., Welling, M.: Modeling relational data with graph convolutional networks. In: European Semantic Web Conference, pp. 593–607 (2018)
21. Bordes, A., Usunier, N., Garciaduran, A., Weston, J., Yakhnenko, O.: Translating embeddings for modeling multi-relational data. In: Neural Information Processing Systems, pp. 2787–2795 (2013)
22. Mikolov, T., Chen, K., Corrado, G., Dean, J.: Efficient estimation of word representations in vector space. Computer Science (2013)
23. Chapelle, O., Metlzer, D., Zhang, Y., Grinspan, P.: Expected reciprocal rank for graded relevance. In: Conference on Information and Knowledge Management, pp. 621–630 (2009)

A Novel Blockchain Based Smart Contract System for eReferral in Healthcare: HealthChain

Shekha Chenthara(✉)(iD), Khandakar Ahmed, Hua Wang, and Frank Whittaker

Institute for Sustainable Industries and Liveable Cities, Victoria University,
Melbourne, Australia
Shekha.Chenthara@live.vu.edu.au,
{Khandakar.Ahmed,Hua.Wang,Frank.Whittaker}@vu.edu.au

Abstract. The privacy of Electronic Health Records is facing a major issue while outsourcing data in the cloud or sharing the records among stakeholders which includes the leakage of private and sensitive information to unauthorized entities. This research mainly focuses on introducing an efficient referral mechanism employing advanced smart contracts for the effective sharing of healthcare records between stakeholders in healthcare industry. This referral system is designed on a patient-centric model and are limited to authorized providers in the healthdata network. This system is built by employing Hyperledger Fabric as the permissioned blockchain utilising Hyperledger composer which visualizes the couchDB and Interplanetary File System as decentralised data storage are combined for efficient and secure big data sharing in healthcare sector.

Keywords: Blockchain · Hyperledger Fabric · Hyperledger composer · IPFS · Smart contracts

1 Introduction

Medical referral is the transition of a patient's treatment upon request from one doctor to another. This framework designs eReferrals so that they can be sent and receive directly between healthcare providers via secure messaging by employing the smart contract functionality in HealthChain. The standard referral management process includes various discrete steps in the communication through faxed paper papers, email and telephone calls. Distributed ledger technology (DLT) also known as blockchain, provides an ideal way to automate the referral process as it provides secure, real-time data exchange between disparate entities, reducing the likelihood of errors, discrepancies and missed connections [28]. Blockchain technology became one of the cutting edge solutions that revolutionizes the healthcare industry by facilitating the secure and efficient sharing of health records among the stakeholders [19]. Besides, it also uses scripting technology called smart contracts that comprises the application logic of the system. Being immutable, trustless, decentralised and distributed, blockchain

© Springer Nature Switzerland AG 2020
Z. Huang et al. (Eds.): HIS 2020, LNCS 12435, pp. 91–102, 2020.
https://doi.org/10.1007/978-3-030-61951-0_9

technology offers wide opportunities for combating fraud, reducing operational costs, optimising processes, eliminating duplication of work and improving transparency in the health care industry.

Nowadays, theft of Electronic Health Records (EHRs) is becoming increasingly pervasive while sharing data due to the poor security and policy enforcement mechanism in the current system [8,22]. This paper introduces a smart healthcare contract system for managing medical data and streamlining challenging medical procedures. This research builds a patient centric permissioned blockchain namely HealthChain built on Hyperledger Fabric by utilizing Hyperledger composer as the rest server API. To avoid failure in third party servers, this work also presents a secure and efficient decentralised platform viz Interplanetary File System for secure data storage. This paper also aims to demonstrate the future use of blockchain in healthcare, and to demonstrate the challenges and possible directions of blockchain technology.

1.1   Contribution

• The main contribution to this research is to provide a Distributed Ledger Technology Smart contract system for efficient eRefferal in medicare as shown in Fig. 3. This work created smart contracts for various medical workflows, and then data access permissions are managed by patient in the healthcare ecosystem shown in Fig. 4.
• This research builds a patient-centeric health chain framework in which patients will have complete control over their medical records maintaining e-health data security, privacy, scalability, and integrity. The HealthChain framework is based on Hyperledger fabric, a permissioned distributed ledger solutions using Hyperledger Composer and stores encrypted EHRs in the InterPlanetary File System (IPFS) to build this private health chain network.
• Another contribution to this research is the introduction of an efficient cryptographical algorithm for data access and storage between stakeholders. This research aims on the scalability of the healthrecords by storing the hash of health records on the chain to maintain the overall efficiency of the blockchain, and actual huge data are stored off the chain storage framework in IPFS, the decentralised storage. Moreover, the data at rest is encrypted by an efficient algorithm based on public key encryption standards.

The rest of the paper is organised as follows: Sect. 2 discusses the blockchain related work in healthcare, Sect. 3 presents the Architecture of the proposed Framework, Sect. 4 presents implementation and simulation results and Sect. 5 as conclusion.

2   Related Works

This section describes the related works on e-health systems using blockchain technology. MedRec is the first permissionless working prototype in healthcare using the Ethereum smart contract functionality for the intelligent representation of medical records stored in individual nodes in the network [10]. Though there is no single point of failure, it fails to achieve scalabiity issues [3,24]. Ancile [9] and Medrec have issues with scalability, which can be overcome by using IPFS via the secure storage offchain rather than the chains itself. Dubovitskaya presented a secure data sharing block chain based on oncology that utilises local database and cloud infrastructure for the storage of encrypted patient data [12,26]. A novel patient-centered architecture has been proposed for fine-grained and flexible data access control using ABE to encrypt EHR data [15,16]. A pemissioned blockchain implementation namely QuorumChain, where only a few users or nodes can vote to add information or blocks to their chain by means of an intelligent contract, thus reducing the complexity of the voting process [18]. Another framework that utilizes hyperledger fabric as the blockchain mechanism for healthcare data sharing that employs mining incentives for providers to access records and also involves a certification authority that oversees every healthcare services [5].

Compared to the current model of sharing of health information, patients choose to use blockchain-enabled applications because of their decentralised data storage characteristics, anonymity, data protection and access control of their EMR and EHR data [13,14,21,25]. In addition, the use of blockchain can expand the current Personal Health Record (PHR) data management system to combine event-driven smart contracts to support transactional services such as repeat prescription, booking appointments, and requests for referrals [14]. And others have deployed a blockchain-enabled decentralised app (DApp) and platform to tackle interoperability issues in health care facilities, allowing patients to use the DApp to exchange their clinical details as the basis for remote support decision-making [9,27]. Nevertheless, most of the clinical environments lack real-world use cases. Most of the existing approaches do not guarantee all the essential requirements for EHRs, such as data privacy, security, secure storage, effective access control, scalability and interoperability [21,23].

Table 1. Comparative study on existing techniques with proposed work

Reference	Security	Privacy	Integrity	Mining	Scalability
Yue [24]	✓	✓	✗	✗	✗
MedRec [3]	✓	✓	✓	✓	✗
Dub [12]	✓	✓	✗	✗	✗
Li [16]	✓	✓	✗	✗	✗
Morgan [18]	✓	✓	✗	✗	✗
Chen [5]	✓	✓	✗	✓	✗
Base paper	✓	✓	✓	✗	✓

Our research work unravels most of the current e-health challenges by using a permissioned Blockchain platform through the use of Practical Byzantine Fault Tolerance (PBFT) as a Consensus to allow data sharing in a decentralised fashion through IPFS by maintaining effective patient privacy, confidentiality and health record integrity [7]. From the comparative study in Table 1, it is clear that the proposed framework resolves most of the issues with the existing techniques and offers a foolproof solution to e-healthdata implementations.

3   Proposed Architecture

The overview of workflow is portrayed in Fig. 1. This framework comprises of stakeholders or participants, Angular 4 application, Fabric SDK, Hyperledger Composer, Hyperledger Fabric, Chaincode, CouchDB and IPFS. Angular 4 is the Front end of the DApp (decentralized application) framework that connects with Composer Rest server which exposes and visualize the state database, couchDB. This application consists of four types of users namely Doctors, Patients, Chemist and Receptionist with n participants for each user. The Fabric-CA provides key public certificates for all n applicants, including patients, doctors, receptionists and pharmacists. The Membership Service Provider abstracts all cryptographic mechanisms such as identity validation, signature generation and verification, certificate issuance and validation protocols and healthchain user authentication. Users can interact with the main application via Angular 4 user interface. User can send and invoke query via Fabric SDK. SDK will verify the global state of the blockchain and submit a query to the blockchain via composer restful service-based API. HealthChain will also send the request for consensus to other peers [17]. Transaction will be submitted to the blockchain after successful consensus, and the subsequent key-value pair will be created or modified according to request. The REST API is used to get the actual state of the couchDB chain database in which the angular frame retrieves data through GET calls to the Rest API of the composer [1,11].

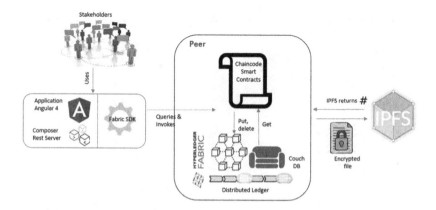

Fig. 1. Overview of workflow in HealthChain network

Hyperledger Fabric is the underlying permissioned blockchain technology for the distributed ledger solutions that supports the building of chaincodes known as smart contracts written in Go, Node.js to validate medical data entries and transactions in the health data network [2]. This organization has three peer nodes and an ordering node with a single public channel for registering the network participants. Application communicates with peer nodes which invokes smart contracts to update the ledger. Whenever a user logs in the application, the credentials are verified from REST API at back end. With every query, the application passes the user credentials on the REST API and gets to know the person's identity that contributes to the first layer of security. When the actual data is retrieved from the blockchain, blockchain checks the user identity in the state database via the REST API when the user has been created on the blockchain that forms the second layer of security. IPFS is used in this research as an off-chain database for the storage and encoding of infinite healthcare records using a public key encryption before storage and hash of the records will be stored in couch database that exposes the blockchain of our framework [4] Any communications with medical records are recorded as network transactions and only the parties participating in the transaction will see the behaviour of the transaction.

This research work uses smart contracts that encompass the application logic of the system for EHR transactions particularly for ereferral between clinicians, data transmission, access management, request handling such as update medical records, update ownerships etc. Smart contracts will be executed during user interaction to identify request, validate request, secure clinician interaction, for granting access permissions and update permissions for medical records.

3.1 Transaction Workflow in Hyperledger Fabric Employing Smart Contracts

This section discusses the work flow of eReferral in HealthChain framework. Assuming the stakeholders are registered participants in the healthchain, initially the Doctor (Clinician) and the patient login with their credentials to the permissioned blockchain. The authorized Doctor (Clinician or General practitioner) checks whether referral is required and refers to specialist practitioner in that scenario with associated patient details. Figure 2 portrays the cryptographical process of adding and sharing records in the proposed system. The referral is performed via employing the eReferral smart contract functionality as shown in Fig. 3 and the access will be given in compliance with the access control rules shown in Fig. 4. The referred doctor can access the EHR for a particular session once the patient approves access to his/her patient details. The specialist reviews and makes updates if required and uploads the record to IPFS after encrypting the record with associated session key [6]. In addition, the specialist on submit lose access to the patient details and the session expires on task completion.

However, if the patient is not a referral case, the doctor performs the normal patient assessment, form a diagnosis, administer patient care and uploads the test result to IPFS. The IPFS return a hash for each transaction and stores the value in couchDB. This research employs Public key encryption for securing data in the off chain database IPFS [20]. Figure 5 shows the step wise explanation of adding records to the HealthChain by referred clinician.

Fig. 2. Cryptographical process of adding records to HealthChain

4   Implementation Results

4.1   Illustration of EHR Access in HealthChain

This approach starts with the assumption that the patient and the clinician have formed an authorised relationship to update health records. The process of adding medical records by clinician to the database is employed via internal encryption mechanism as shown in Fig. 2. The referred clinician will be able to view the referred details by the General Practitioner. Whenever a new patient record will be added or modified by the referred clinician, the system creates a composite view, P_{Cv_i} of the data that can be accessible to the clinician C_i alternately sharing the whole data. Composite view P_{Cv_i} is the attribute set of the stored medical record P_{EHR_i} that the system creates on permissioned user request without sharing the complete patient record as shown in Eq. (1).

The system further generates a session key S_k shared by patient and the referred clinician for a distinct session. The system then sends the encrypted session key S_k to the patient as $E_{P_{pk_i}}(S_k)$ and clinician as $E_{C_{pk_i}}(S_k)$ by encrypting using respective public keys of the patient P_{Pk_i} and clinician C_{Pk_i} for a distinct session. The composite view P_{Cv_i} will also be encrypted with session key S_k as $E_{S_k}(P_{Cv_i})$ and stores in IPFS. Inaddition, the system sends encrypted composite view i.e. $E_{S_k}(P_{Cv_i})$ to the clinician. Now, clinician decrypts the session key with his private key and decrypts the composite view with the session key. If there are any updates, clinician updates P_{Cv_i} as UP_{Cv_i}, resolves the case, encrypts with the session key and uploads UP_{Cv_i} to IPFS as $E_{S_k}(UP_{Cv_i})$. Clinician also

Algorithm 1: Smart Contract for Patient records

Assign Roles to stakeholders
function Define Roles (New role, New Account)
    Add new role and new account in Healthchain
    access based on access control permission rules
end function
function create medical record (contains asset variables to create record)
    If (msg.sender_id == GPdoctor_id) then
        Create medical record to patient's record
        IPFS $\longleftarrow$ E_{S_k} (UP_{Cv_i}) /*Encrypts updated composite view with Clinician's session key
        return hash #   /* Return hash value to CouchDB, blockchain
    else Abort session
    end if
end function
function create patient referral record (contains asset variables for referral)
    If (GPdoctor_id == doctor_id && referdoctor_id== doctor_id) then
    if (patient_id ==true && record_id==true)
        return record from specific patient_id
        create patient referral record
        update data to particular patient's record
        return hash #
    else Abort session
    end if
    end if
end function
function view patient record (patient id)
    if (msg.sender_id == doctor || patient) then
    if (doctor_id == true && patient_id==true) then
        return patient record
    else Abort session
    end if
    end if
    end function

function update patient record (contains asset variables to update patient record)
    if (msg.sender_id == doctor) then
    if (doctor_id == true && patient_id==true) then
        P_{EHR_i} $\longleftarrow$ [(D_{PPri}($E_{P_{Pki}}$ (P_{EHR_i}))) + $E_{P_{Pki}}$(UP_{Cv_i})]   /* Store updated medical record to IPFS
        IPFS $\longleftarrow$ UP_{EHR_i} update patient record
        return hash #
    else return fail
    end if
    else Abort session
    end if
end function

Fig. 3. Snippet of smart contract for eReferral

sends patient referral $E_{S_k}(RP_{Cv_i})$ if required to the specialist practitioner. Specialist decrypts the associated session key S_k, decrypts and reads the (RP_{Cv_i}), request Patient P_i for more patient details. The process of generating composite view repeats and the specialist updates P_{Cv_i} as UP_{Cv_i} encrypts with session key $E_{S_k}(UP_{Cv_i})$, stores in IPFS and resolves the case. On clinicians' record update, the system decrypts the encrypted record i.e. $E_{P_{Pk_i}}(P_{EHR_i})$ using patients'

```
/* Access control rules for EHR-network*/
•   rule DoctorCanReadPatient {
        description: "Allow doctor read access to all granted patients"
        participant(p): "org.ehr.healthchain.Doctor"
        operation: READ
        resource(r): "org.ehr.healthchain.Patient"
        condition: (r.authorized && r.authorized.indexOf(r.getIdentifier()))>-1)
        action: ALLOW }
•   rule DoctorCanUpdateEHR {
        description: "Allow doctor update access to all granted patients"
        participant(p): "org.ehr.healthchain.Doctor"
        operation: CREATE,UPDATE
        resource(r): "org.ehr.healthchain.Patient"
        transaction(tx):"org.ehr.healthchain.UpdateRecord"
        condition: (r.authorized && r.authorized.indexOf(p.getIdentifier()))>-1)
        action: ALLOW }
•   rule PatientCanReadEHR {
        description: "Allow patient read access to his/her own records"
        participant(p): "org.ehr.healthchain.Patient"
        operation: READ
        resource(r): "org.ehr.healthchain.Medical_Record"
        condition: (p.PatientId == r.PatientId)
        action: ALLOW }
```

Fig. 4. A snippet of Access control definition in HealthChain framework

private key and also decrypts the encrypted updated composite view from the IPFS i.e. $E_{S_k}(UP_{Cv_i})$ using the session key as shown in Eq. (2). Finally, the system commits the updates to the original record and encrypts the original record P_{EHR_i} as $E_{P_{Pk_i}}(P_{EHR_i})$ before uploading to IPFS as shown in Eq. (3). The session key S_k for each session expires and the composite view P_{Cv_i} will be deleted upon session completion. The transactions eventuated on clinician access and record updates will be hashed by employing smart contracts and added to the healthchain. Figure 6 shows the process of referring health records for Doctor Referral by employing essential attributes in the HealthChain.

$$P_{Cv_i} \subseteq P_{EHR_i} \tag{1}$$

$$P_{Cv_i} = (D_{P_{Pr_i}}(E_{P_{Pk_i}}(P_{EHR_i}))) \tag{2}$$

$$P_{EHR_i} = [(D_{P_{Pr_i}}(E_{P_{Pk_i}}(P_{EHR_i}))) + (E_{P_{Pk_i}}(U_{P_{Cv_i}}))] \tag{3}$$

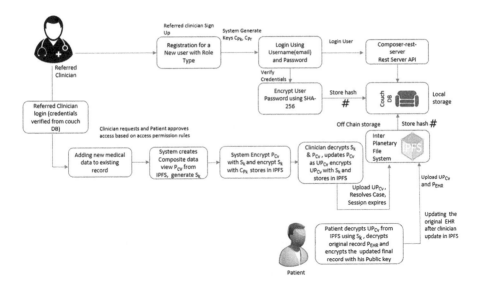

Fig. 5. Illustration of adding records to HealthChain

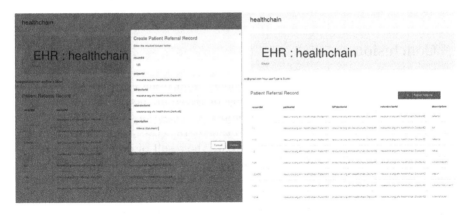

Fig. 6. Illustration of referring records in HealthChain

4.2 Demonstration of IPFS Scalability in HealthChain Framework

Figure 7 demonstrates the scalability of IPFS using both the image data and pdf document data with a size comparison upto 100 MB size. The results are obtained from transaction execution of 5 users concurrently upload and download the data in IPFS. Considering the system requirements, for a 100 MB file, the system takes an average time of 65 s to upload the data to IPFS and downloading in an average time of 105 s for a data (.pdf) document. In the case of an image data, it takes an average of 65 s to upload and 80 s to download for a 100 MB image file.

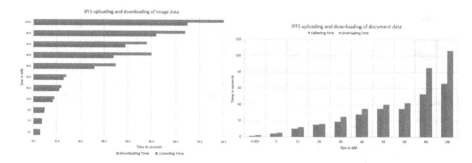

Fig. 7. Uploading and Downloading time comparison of image data and pdf document in IPFS

The simulation is conducted in a virtual machine environment (Ubuntu 16.04 LTS) and the PC has the following configurations.
• 2 Core CPU (Intel Core i5 2.5 GHz (Turbo Boost up to 2.7 GHz) with 3 MB shared L3 cache
• 5 GB Memory
• 1 Gbit/s network • 30 GB SSD

5   Conclusion

In this research work, a permissioned Block chain framework has been implemented for secure data storage and access of electronic health records utilizing Hyperledger fabric and Hyperledger composer. This work created smart contracts for various medical workflows, and then data access permissions are managed by patient in the healthcare ecosystem. The result of prototype implementation and analysis proves that the approach is a tamper resistant mechanism as information will be stored as hash values for every healthcare transactions in the blockchain. Moreover, it has enormous potential to ensure privacy, security, integrity, confidentiality and scalability of the e-health information. This research also explores technology framework and business processes for blockchain applications.

References

1. Anderson, J.C., Lehnardt, J., Slater, N.: CouchDB: The Definitive Guide: Time To Relax. O'Reilly Media Inc., Sebastopol (2010)
2. Androulaki, E., et al.: Hyperledger fabric: a distributed operating system for permissioned blockchains. In: Proceedings of the Thirteenth EuroSys Conference, p. 30. ACM (2018)
3. Azaria, A., Ekblaw, A., Vieira, T., Lippman, A.: MedRec: using blockchain for medical data access and permission management. In: 2016 2nd International Conference on Open and Big Data (OBD), pp. 25–30. IEEE (2016)

4. Benet, J.: IPFS-content addressed, versioned, P2P file system. arXiv preprint arXiv:1407.3561 (2014)
5. Chen, J., Ma, X., Du, M., Wang, Z.: A blockchain application for medical information sharing. In: 2018 IEEE International Symposium on Innovation and Entrepreneurship (TEMS-ISIE), pp. 1–7. IEEE (2018)
6. Cheng, K., et al.: Secure k-NN query on encrypted cloud data with multiple keys. IEEE Trans. Big Data **PP**, 1 (2017)
7. Chenthara, S., Ahmed, K., Wang, H., Whittaker, F.: Security and privacy-preserving challenges of e-health solutions in cloud computing. IEEE Access **7**, 74361–74382 (2019)
8. Chenthara, S., Wang, H., Ahmed, K. Security and privacy in big data environment. In: Sakr, S., Zomaya, A. (eds.) Encyclopedia of Big Data Technologies. Springer, Cham (2018). https://doi.org/10.1007/978-3-319-63962-8_245-1
9. Dagher, G.G., Mohler, J., Milojkovic, M., Marella, P.B.: Ancile: privacy-preserving framework for access control and interoperability of electronic health records using blockchain technology. Sustain. Cities Soc. **39**, 283–297 (2018)
10. Dannen, C.: Introducing Ethereum and Solidity. Apress, Berkeley (2017). https://doi.org/10.1007/978-1-4842-2535-6_9
11. Dhillon, V., Metcalf, D., Hooper, M.: The hyperledger project. Blockchain Enabled Applications, pp. 139–149. Apress, Berkeley (2017). https://doi.org/10.1007/978-1-4842-3081-7_10
12. Dubovitskaya, A., Xu, Z., Ryu, S., Schumacher, M., Wang, F.: Secure and trustable electronic medical records sharing using blockchain. In: AMIA Annual Symposium Proceedings, vol. 2017, p. 650. American Medical Informatics Association (2017)
13. Esmaeilzadeh, P., Mirzaei, T.: The potential of blockchain technology for health information exchange: experimental study from patients' perspectives. J. Med. Internet Res. **21**(6), e14184 (2019)
14. Leeming, G., Cunningham, J., Ainsworth, J.: A ledger of me: personalizing healthcare using blockchain technology. Front. Med. **6**, 171 (2019)
15. Li, M., Sun, X., Wang, H., Zhang, Y., Zhang, J.: Privacy-aware access control with trust management in web service. World Wide Web **14**(4), 407–430 (2011). https://doi.org/10.1007/s11280-011-0114-8
16. Li, M., Yu, S., Zheng, Y., Ren, K., Lou, W.: Scalable and secure sharing of personal health records in cloud computing using attribute-based encryption. IEEE Trans. Parallel Distrib. Syst. **24**(1), 131–143 (2012)
17. Mingxiao, D., Xiaofeng, M., Zhe, Z., Xiangwei, W., Qijun, C.: A review on consensus algorithm of blockchain. In: 2017 IEEE International Conference on Systems, Man, and Cybernetics (SMC), pp. 2567–2572. IEEE (2017)
18. Morgan, J.: Quorum Whitepaper. JP Morgan Chase, New York (2016)
19. Nakamoto, S.: Bitcoin: a peer-to-peer electronic cash system. Technical report, Manubot (2019)
20. Salomaa, A.: Public-Key Cryptography. Springer, Heidelberg (2013)
21. Vimalachandran, P., Wang, H., Zhang, Y., Zhuo, G.: The Australian PCEHR system: ensuring privacy and security through an improved access control mechanism. arXiv preprint arXiv:1710.07778 (2017)
22. Wang, H., Wang, Y., Taleb, T., Jiang, X.: Special issue on security and privacy in network computing. World Wide Web **23**(2), 951–957 (2020). https://doi.org/10.1007/s11280-019-00704-x
23. Wang, H., Zhang, Z., Taleb, T.: Special issue on security and privacy of IoT. World Wide Web **21**(1), 1–6 (2018). https://doi.org/10.1007/s11280-017-0490-9

24. Yue, X., Wang, H., Jin, D., Li, M., Jiang, W.: Healthcare data gateways: found healthcare intelligence on blockchain with novel privacy risk control. J. Med. Syst. **40**(10), 218 (2016). https://doi.org/10.1007/s10916-016-0574-6

25. Zhang, J., et al.: On efficient and robust anonymization for privacy protection on massive streaming categorical information. IEEE Trans. Dependable Secure Comput. **14**(5), 507–520 (2015)

26. Zhang, J., Tao, X., Wang, H.: Outlier detection from large distributed databases. World Wide Web **17**(4), 539–568 (2013). https://doi.org/10.1007/s11280-013-0218-4

27. Zhang, P., White, J., Schmidt, D.C., Lenz, G.: Applying software patterns to address interoperability in blockchain-based healthcare apps. arXiv preprint arXiv:1706.03700 (2017)

28. Zyskind, G., Nathan, O., et al.: Decentralizing privacy: using blockchain to protect personal data. In: 2015 IEEE Security and Privacy Workshops, pp. 180–184. IEEE (2015)

Telemedicine System with Elements of Artificial Intelligence for Health Monitoring During COVID-19 Pandemic

Sergei Shinkariov[1]([✉]), Boris Zingerman[2] [ID], Irina Kargalskaya[3], Arkadii Nozik[2], Inna Fistul[2], Lev Evelson[4] [ID], Alexandra Kremenetskaya[5], Le Sun[6] [ID], Jun Xu[7] [ID], Olga Kremenetskaya[8] [ID], and Nikita Shklovskiy-Kordi[5] [ID]

[1] Lipetsk Oncological Hospital,
1-e, Admiral Makarov Street, Lipetsk 398005, Russian Federation
seam20062@yandex.ru
[2] TelePat LLC, Moscow 115533, Russian Federation
boriszing@gmail.com
[3] The Committee "Patient-oriented Telemedicine", Moscow 109390, Russian Federation
kargalska@yandex.ru
[4] Innovation Scientific Center of Information and Remote Technologies,
Bryansk 645325, Russian Federation
levelmoscow@mail.ru
[5] National Research Center for Hematology, Moscow 125167, Russian Federation
nikitashk@gmail.com
[6] School of Computer, Nanjing University of Information Science
and Technology, Nanjing 210044, China
sunle2009@gmail.com
[7] School of Automation, Nanjing University of Information Science
and Technology, Nanjing 210044, China
xujung@gmail.com
[8] Center for Theoretical Problems of Physico-chemical Pharmacology, Moscow 119991, Russia

Abstract. The ONCONET telemedicine system, intended for remote monitoring of the health status of cancer patients, is presented. The interactive part in asynchronous mode provides virtual contacts in form of medical records: patient's questions, doctor's answers, questionnaires filled up by patients. On-line video conversations are possible in emergency. The patient can add any medical documents in his health monitoring record. The analytical subsystem, using artificial intelligence elements, reveals signs of alarm situations in patient messages automatically. The subsystem estimates necessity to demand attention of the doctor or emergency services. Special questionnaires devoted to COVID-19 had been developed. All the data can be represented in integrated form on common-time scale graphs and colored diagrams ("heat maps") reflecting health statement of a patient. There are video teaching cases and medical information materials particularly connected to COVID-19. The System collects, organizes and saves personal medical information according with personal electronic medical case history structure. The system had been tested in 22 medical organizations in Russia. Ways of further perspective research and development of the system are discussed.

© Springer Nature Switzerland AG 2020
Z. Huang et al. (Eds.): HIS 2020, LNCS 12435, pp. 103–110, 2020.
https://doi.org/10.1007/978-3-030-61951-0_10

Keywords: Telemedicine · COVID-19 · Artificial intelligence · Health remote monitoring · Cancer patients · Medical data

1 Introduction

The cancer patients usually suffer both the disease itself and the side effects of anticancer treatment. They also make up the particular risk group in pandemic. Differences in the individual course of the disease demand permanent supervision. Periodical visits don't provide sufficient communication between doctor and patient [1]. Problems caused by the pandemic greatly exacerbate the negative circumstances inherent in cancer. Besides of problems connected with quarantine, if any cancer patient had the COVID-19 disease (even in light form) any unknown consequences could take place.

Telemedicine can help to solve the problems. It can provide effective interaction between patient and doctor [2]. Now many countries are actively using computer technology involving patient's active report about the health condition and transmitting it to the health care providers [3]. Permanent communication between a patient and a doctor can have the same or even more effectiveness as some very expensive drug. However, until we invest much more money into development of drugs than into remote monitoring of health [4].

The telemedicine system devoted to remote monitoring health is presented in this article. The study had been started for cancer patients since the beginning of 2019. The software had been developed before Covid-19. In the context of the COVID-19 pandemic, additional components were developed and inserted into the system. They allow detect situations with high risk of the disease and inform the patient and the doctor about it.

Below short description of proposed system and results of the first stage implementation in Russia are presented. It is the first similar experiment in Russia.

Perspectives of further development of the system and opportunities of international co-operation are contained in "Discussion" and "Conclusions".

2 Materials and Methods

A specialized telemedicine platform ONCONET [5] had been developed and used for remote monitoring cancer patients. ONCONET is a cloud service. It is granted to medical organizations as a complete fully-finished service. A medical organization registers on the platform and receives a virtual management office. The platform consists of information part, interactive service, technological and analytical modules. The information part of the platform includes a library for patients (information materials and video teaching cases) and a doctor's library containing medical materials. The interactive part consists of personal doctors' and patient's virtual offices. The technological subsystem has a questionnaire service and a special design tool that can help to create quickly new monitoring schemes. An important element of the system is an analytical module that provides visualization of monitored data and classifies patients according to the severity and urgency of the necessary attention. The colored diagram picture («heat map») and monitoring curves can be the results of the visualization approach. Data stored in different formats (text, tables, roentgenograms, microphotographs, videos etc.) can be displayed in an integrated fashion on the same screen. Automatic matching of the data to stored timetables and established protocols for diagnostics and treatment correction takes place.

The above specialized questionnaires were developed by oncologists for remote assessment of the patient's condition. One of the main problems in compiling questionnaires was to convert the evaluation descriptions into a language understandable for the patient. Patient communities and large groups of patients were involved in the development of questionnaires to assess their comprehension and testing. The ONCONET has a special toolkit that allows add new and perfect existing questionnaires. In order to control important symptoms and side effects typical for specific treatments and cancer patient conditions, many profile questionnaires had been created. In the beginning of the COVID-19 pandemic, a questionnaire on the symptoms of COVID-19 was added to the system.

The system automatically performs data validation and notifies a user when selected parameters are beyond acceptable ranges or when the timetable set by the protocol is not followed. These software features permit to monitor and correct individual actions taken by patient. Attention of physicians and staff is prompted by a color indicator. Depending on the type of situation, the system suggests several different responses. They include cancellation of an inquiry or change in assigned medication or dosage. Notification about specific situation is automatically dispatched to the address of the individual in charge of the protocol management, attending physician and other personnel specified by the user. The problem is detected in real-time and the system facilitates collective decisions for corrective action to avoid damage. So, there is a convenient tool for entering available information about a patient. It provides easy access to primary data and allows generate multimedia presentation of a patient's record in the common time-line axis format. The system links multi-format medical data forming a recognizable image of disease and allows its real-time evaluation [6].

A patient fills up the questionnaires appointed by the doctor. Each possible answer to a question (the patient can choose from the list) is ranked by the severity of the condition and receives a color (from green to red). The patient sees the "color of the answer". All the patient's answers are displayed in a table where the observation dates are vertically, and horizontally - controlled symptoms. Accordingly, each cell indicates the "color of the condition" chosen by the patient. It allows the doctor to grasp the yellow and red fragments with a "one look" and understand the problems. In addition to heat maps there are graphs of measured quantitative parameters (temperature, pressure, weight, etc.) and marks about taken drugs, placed in columns corresponding to the date. It also allows the doctor to compare quickly the commitment to treatment with dynamics of the parameters and color display of well-being. Each question has assigned weight, and each answer is evaluated by numerical value. These weighted values are summarized and form an integral assessment of the patient, after which patients are ranked and "painted" (from green to red). The doctor receives a ranked list of patients, in which the heaviest cases are shown at the top and highlighted by color. In the first level the weight of questions and the ratios of answers can be put by doctors manually. However, in the second level we stipulate for automatic estimation of the patient health statement by software based on artificial intelligence methods [7]. Artificial neural network trained on collected results of the system work would analyze the questionnaires.

3 Procedures

The questionnaires include questions and answers in the table in one of following 2 forms. a) Descriptive response options ranked from the norm (marked green) to critical state (marked red). b) In the form of a score from 0 to 10 (the assessment is chosen by the patient with the help of the engine). The patient receives a questionnaire in which all values have been pre-installed in the value of "norm" (green). But if the patient feels problems for any of the symptoms, he can open the question and specify the value most appropriate to his condition. The questionnaires sent to the patient can also be used to assess his psychological state by the methods developed in clinical psychology. The attending physician prescribes the patient a set of specialized questionnaires and the frequency (schedule) of their referral to the patient. If the patient indicates a dangerous or critical symptom in the questionnaire, he receives at once a notification like as "Call an ambulance immediately!" or "Appeal urgently to the doctor!" etc.

The patient also receives a set of recommendations in the form of references to information materials or training video cases posted in the same portal ONCONET. It is also possible for the patient to attach additional documents (for example, results of laboratory test) and write a message to the doctor describing a symptom or problem not included in the questionnaire. The completed questionnaires are stored on the server and sent to the doctor in the form of a special "heat map" demonstrating the dynamics of symptoms. The questionnaire includes symptoms and the patient has to confirm or not confirm "degree of presence" of them: Nausea – no nausea; Body temperature: normal—not normal, etc. There are principal symptoms: fever, diarrhea, nausea, vomiting, weakness, insomnia, heart palpitations, appetite disorder, shortness of breath and others. Then symptoms connected with control of nervous system take place: vertigo, balance problems, headache, numbness, memory weakness. There is the block of symptoms connected with control of emotional state: anxiety, despondency, irritability, etc. There are also basic for chemotherapy treatment symptoms. They are following: nausea, fever, vomiting, diarrhea, constipation, urination, edema, weakness, bleeding (coagulation disorders), slowing pulse, accelerating pulse, shortness of breath, weakness, etc.

Periodically the doctor receives a summary of all his monitored patients. The summary is ranked on the overall complex severity of the symptoms noted by the patient, so that the doctor can evaluate the condition of the most severe patients more quickly and make appropriate actions. The integrated evaluation algorithm had been developed by the authors. It takes into account the weight strains of various symptoms indicated by oncologists. If necessary, the doctor can directly write an additional recommendation to the patient in the ONCONET system, invite him to a face-to-face or on-line appointment or send a link to additional material with recommendations.

The doctor assigns the schedule of quantitative parameters control (blood pressure, temperature, glucose, weight etc.), as well as reminders of taking drugs, with confirmation of the reception. The patient's indicators and drug treatment marks are also included in the "heat map" in the form of clear graphs showing changes in time.

ONCONET integrates the Patient Library including 1,250 pages of patient-oriented information content and supporting patient teaching cases from leading experts on the specifics and treatments for life and nutrition, rehabilitation and care. It is structured on oncological diagnoses (21 types of cancers) as well as stages of the patient's life.

After that the COVID-19 pandemic started, 13 video teaching cases were added to the information part of the system. They concerned the life conditions of oncological patients under the COVID-19 quarantine situation (it was done together with prof. N.V.Zhukov). Those cases present many important recommendations such as: how to control breath under the COVID-19, prophylaxis peculiarities for patients with cancer, psychological aspects during pandemic, etc.

4 Results

The system was tested during 2018–2020 in 22 medical organizations in 10 regions of Russia. 174 doctors and 382 patients participated in the testing of the system (at the time of publication). A total of 21,936 questionnaires (an average of 57 per patient) had been filled out by patients and 193,248 symptoms had been reported.

The largest number of patients has been tested in the Lipetsk Regional Oncology Clinic during implementation of the ONCONET system since 2019 year. 61 doctors took part in the project. They provided remote monitoring of 206 patients. The study involved patients with breast cancer, lung, cervical and ovarian cancer, prostate, thyroid, kidney, stomach and colorectal cancer, lymphoma. The comparison group of patients not involved into remote monitoring (209 persons) had been selected. Similarity in composition of diagnoses and stages of diseases and treatment regimens, as well as on the sex-age composition was observed. According to the results of the surveys, patients indicated: the "normal" condition of the symptom in 89.9% of cases (scattering from 8% to 100% for different symptoms); a state with certain deviations from the "norm" in 8% of cases (spread from 0% to 92%); severe and critical states in 2.1% of cases (spread from 0% to 47%).

The feedback on the system received from doctors and patients was positive. The difference in the frequency of treatment complications for patients participating in ONC0NET and the comparison group had been found following:

- hematological complications - 32% in the ONCONET group vs. 42% in the comparison group);
- disruptions in the gastrointestinal (0.7% vs. 2.6%);
- neurotoxic manifestations are practically reduced to a minimum (0% vs. 3.2%);
- nephrotoxicity (3.6% vs. 13.6%);
- skin complications were reduced to zero (0% in the group ONCONET vs. 1.6% in the control) Thanks to the skin photos attached to the questionnaires, the phenomenon was detected at an early stage.

The postponements of the next chemotherapy course (very important parameter) were by 1,5 times less (7.9% for ONCONET vs. 11.9% in the comparison group).

5   Discussion

The creation and implementation of the ONCONET system was carried out at the request of the patients of the Association of Cancer Patients "Hello". The deep interest of chronic patients in such medical service was confirmed by a survey conducted by us for annual All-Russian Patient Congress. The assessment of the demand for various remote services was done (300 participants in survey). 77% believed that the permanent channel of communication with the doctor would help to reduce fear and not to feel "one-on-one" with the disease and consequences of toxic treatment. So, course of treatment would be not interrupted due to uncontrolled complications.

This study is the first experiment of proving the effectiveness of remote monitoring in oncology in Russia. Now we are starting a prospective randomized study with the second level of the system including artificial intelligence elements to assess the impact of remote monitoring on the quality of life of patients.

The Covid-19 pandemic has become a catalyst to accelerate development and application of remote technologies in medicine. During the pandemic period, a module was integrated into the ONCONET system, which automates the process of detecting COVID-19 cases (more exactly – detecting high risk of such cases) among cancer patients undergoing treatment and observation. It should reduce the number of visits of sick patients to hospitals and help to reduce the risk of infection for other patients and doctors. The proposed system would allow unloading round-the-clock and day hospitals. Besides, it can be useful for mass tests of the vaccine.

It is planned to involve in processing of answers in questionnaires a set of intelligent agents, computer modules that allow not only automatically process the answers, but also generate questions, conduct a clarifying dialogue with the patient, structure the answers received in such way.

For perspective, it is intended to supplement the proposed telemedicine system with a component for analysis of depersonalized data. It would be possible to analyze the state of health of the population, including psychological state, to see the dynamics of processes, to compare the situation in different regions, search new regularities in treatment methods, etc. In our view, this way of developing the system meets the urgent needs of medicine and society as a whole. The patient would become the main person in the management of his health. Medical care should take into account the individual characteristics as much as possible. This approach also provides an opportunity to investigate the patterns of the COVID-19 and other similar infections. An important aspect is ability to identify anomalies that deviate greatly from the found patterns [8]. Another promising area of development for the proposed telemedicine system is the widespread use of IoT (Internet of Things) technology in conjunction with artificial intelligence methods to automatically analyze the data coming from sensors [9]. Application of artificial intelligence in the classification of histopathological images and the analysis of their evolution in time [10] can also be an important useful direction of research and of evolution of the proposed system.

6 Conclusions

The study demonstrates the great benefit of remote monitoring of cancer patients implemented by the proposed telemedicine system. The high medical effectiveness takes place especially in conditions of the pandemic and forced self-isolation. Remote monitoring would also improve the effectiveness of the treatment and also reduce the loading of the oncology services. It is claimed much by cancer patients. It significantly improves their quality of life, satisfaction from the treatment, loyalty to the ongoing treatment and recommended lifestyle. In the long term, the broad implementation of the proposed telemedicine system, its development in the direction of collecting, consolidating and analyzing depersonalized medical data, more active involving artificial intelligence in monitoring is appropriate and relevant. It's a step towards individual patient-oriented medicine. On the other hand, it would provide an opportunity to analyze the situation with public health (including, patterns of the spread of the pandemic and its consequences) and to make adequate management decisions. The use of the System in mass tests of the vaccines is under consideration.

Acknowledgments. This work was supported in part by the Ministry of Science and Higher Education of the Russian Federation (AAAA-A18-118012390247-0) and by RFBR grant (project № 19-07-01235).

References

1. Benze, G., et al.: PROutine: a feasibility study assessing surveillance of electronic patient reported outcomes and adherence via smartphone app in advanced cancer. Ann. Palliat. Med. https://doi.org/10.21037/apm.2017.07.05
2. Holch, P., et al.: Development of an integrated electronic platform for patient self-report and management of adverse events during cancer treatment. Ann. Oncol. **28**(9), 2305–2311 (2017)
3. Adler-Milstein, J., Longhurst, K.: Assessment of patient use of a new approach to access health record data among 12 US health systems JAMA Netw. Open https://doi.org/10.1001/jamanetworkopen.2019.9544
4. https://www.healio.com/hematology-oncology/practice-management/news/online/%7Be8c2c242-8bf0-4b8a-88a6-e5f7d9e8a2f0%7D/online-symptom-reporting-may-help-patients-with-cancer-live-longer
5. TelePat Homepage. https://telepat.online/
6. Shklovskiy-Kordi, N., Borodin, R., Zingerman, B., Shifrin, M., Kremenetskaya, O., Vorobiev, A.: Web-service medical messenger – intelligent algorithm of remote counseling. In: Siuly, S., Lee, I., Huang, Z., Zhou, R., Wang, H., Xiang, W. (eds.) HIS 2018. LNCS, vol. 11148, pp. 193–197. Springer, Cham (2018). https://doi.org/10.1007/978-3-030-01078-2_18
7. Shklovskiy-Kordi, N., Shifrin, M., Zingerman, B., Vorobiev, A.: Some directions of medical informatics in Russia. In: Siuly, S., Huang, Z., Aickelin, U., Zhou, R., Wang, H., Zhang, Y., Klimenko, S. (eds.) HIS 2017. LNCS, vol. 10594, pp. 22–31. Springer, Cham (2017). https://doi.org/10.1007/978-3-319-69182-4_3
8. Ma, J., Sun, L., Wang, H., et al.: Supervised anomaly detection in uncertain pseudoperiodic data streams. ACM Trans. Internet Technol. **16**(1) (2016). Article no. 4, 20 pages

9. Sun, L., Dong, H., Liu, A.X.: Aggregation functions considering criteria interrelationships in fuzzy multi-criteria decision making: state-of-the-art. IEEE Access **6**, 68104–68136 (2018)
10. Xu, J., Luo, X., Wang, G., et al.: A deep convolutional neural network for segmenting and classifying epithelial and stromal regions in histopathological images. Neurocomputing **191**, 214–223 (2016)

Genetic Interpretation System for Screening Monogenic Disorders Carriers

Jitao Yang$^{(\boxtimes)}$ and Bin Li

School of Information Science, Beijing Language and Culture University,
Beijing 100083, China
yangjitao@blcu.edu.cn

Abstract. Monogenic disorders are remained a common issue in some countries due to poor environmental factors and unnecessary mutations, more specifically in the rural areas, the monogenic disorder ratio is much more higher because of the consanguineous marriages. According to the OMIM database, there are currently more than 8,000 single-gene diseases identified. Although monogenic diseases are rare, the overall incidence is nearly 1/100. Among various birth defects, the proportion of single gene diseases is as high as 22.2%. Among neonatal deaths, 20% of them are caused by recessive genetic diseases, and 80% of patients with recessive genetic diseases have no family genetic history. Screening of carriers of single-gene inherited diseases based on specific populations has been proposed since the 1970s, however the cost was very high and it was impossible to be widely used. In the past few years, with the fast development of genetic testing technology and clinical applications, more and more single-gene associated diseases have been discovered, and the cost for screening monogenic disorders has become less and less. In this paper we first introduce the single-gene inherited diseases, then we explain how to identify the single-gene inherited diseases, finally we describe the working mechanism and implementation of the genetic interpretation system for screening monogenic disorders carriers.

Keywords: Monogenic disorders · Genetic testing · Carrier
screening · Genetic interpretation

1 Introduction

It's well known that each person has two sets of DNA, one from father and another from mother. When a sperm carrying father's DNA fertilizes an egg containing mother's DNA, those two sets of DNA combine to make the new sets of DNA, which are the unique hereditary material of the baby. The hereditary material determines person's hair color, height, skin condition, nutrition requirement, sports capability, and even the risk of diseases.

The health conditions caused by gene mutations are called genetic disorders. Some mutations are harmless, some mutations will slightly boost the risk of a

© Springer Nature Switzerland AG 2020
Z. Huang et al. (Eds.): HIS 2020, LNCS 12435, pp. 111–118, 2020.
https://doi.org/10.1007/978-3-030-61951-0_11

health condition, however, some mutations can cause serious diseases or growth and development problems starting at the birth of a baby. In the past few years, with the development of genetic technology and clinical applications, more and more single-gene associated diseases have been discovered [1], according to the OMIM [17,18] database, there are currently more than 8,000 single-gene diseases identified. The single gene associated disorders are defined as monogenic disorders.

Monogenic disorders have been remained a common issue in some countries due to poor environmental factors and unnecessary mutations, more specifically, in the rural areas with high consanguineous marriages will cause more monogenic disorders carries. Although monogenic diseases are rare, the overall incidence is nearly 1/100.

According to the statistic data of the "Report on the Prevention of Birth Defects in China (2012)" [16] released by the Ministry of Health, the incidence of birth defects is common in China, with a birth defect rate of 5.6%, which means nearly 900,000 babies are affected by birth defects each year.

Based on the "global report on birth defects" [4], among various birth defects, the proportion of single gene diseases is as high as 22.2%. Among neonatal deaths, 20% of them are caused by recessive genetic diseases, and 80% of patients with recessive genetic diseases have no family genetic history.

Therefore, it is very important and necessary to have monogenic disease testing before or during pregnancy, so that to block the genetic disease-causing genes to pass on to the next generation, and make the rare diseases even rarer.

In this paper, we first introduce the knowledge of monogenic disorders and carriers, then we describe how to reduce the monogenic disorders through genetic testing/screening, finally we give the design and implementation of a monogenic disorders carriers screening system.

2   Monogenic Disorders and Carriers Screening

2.1   Monogenic Disorders Carriers

Carrier is the person who usually has no symptoms or has only mild symptoms of a disorder, and the person does not know that he/she has the defective gene causing the disorder. This monogenic disorder is recessive, which means the baby carrier must inherits the defective gene either from father or from mother. If both the father and mother are carriers of a monogenic disorder like cystic fibrosis, or Tay-Sachs disease, the child will have 25% chance of having the disease. Because each of the parent has two sets of DNA, since the parent carriers don't actually have the phenotype of disease, that means each patent has a second healthy set of DNA copy, if their baby inherits the healthy copy from one parent or inherits the healthy copies from parents, the baby won't have the disease (although the baby may be a carrier). But 25% is a very high probability.

In monogenic disorders, the mutated genes are inherited and the caused characteristic phenotypes are following the mendelian segregation patterns [2]. The patterns of inheritance can be predicted in both cases of autosomes as well as sex chromosomes. The patterns also describe whether a single copy of a gene is inherited and responsible for causing the disease [3].

Carrier screening is the genetic testing that performed on an individual who does not have any overt phenotype of monogenetic disorder but may have one variant allele within gene associated with recessive genetic disease. Carrier screening provides life-lasting information about an individual's reproductive risk and his/her possibility of having child with genetic disease. Parents are suggested to have genetic disorders testing before or during pregnancy to avoid the genetic disorder to be passed on to the next generation. Carrier screening is recommended for at least one parent, and the testing of the second parent becomes much more necessary if the first parent was tested positive.

Carrier screening was used to identify the persons at high risk of having a child with a serious genetic disorder long time ago, screening of carriers of single-gene inherited diseases for specific populations has been proposed since the 1970s. Initially, screening was recommended for couples at high risk of passing on specific single-gene disease to the next generation, such as the Tay-Sachs Disease (TSD), the TSD carrier screening was the program for childbearing German Jews in the United States and Canada in 1970 [5].

2.2 Carriers Screening with NGS

The emergence of high-throughput sequencing technology has greatly improved the testing efficiency and reduced the cost, so that specific populations and more general populations with no family disease history can be screened for various monogenic diseases before or during pregnancy [6,7].

The next generation sequencing (NGS) was reported to be used for screening genetic diseases carriers by Bell et al. (2011) [8] that, in the research, 104 unrelated individuals were screened for 448 severe recessive genetic diseases, the result showed that each person carried an average of 2.8 pathogenic mutations, which means the necessity of screening the monogenic disorders carriers for general population. The research also demonstrated that the screening of the carriers using NGS technology has the characteristics of cost-effective, which makes the feasibility of having a broad screening of population. Gabriel A. Lazarin et al. (2013) reported the analysis of 108 recessive genetic diseases in 23,453 clinical samples from multiple races, and found that 24% of individuals were carriers of at least one disease, 5.2% of the individuals are carriers of multiple diseases [9].

Azimi M et al. [10] (2015) released NGS clinical verification data for carrier screening, in which, 48,761 clinical samples were tested through NGS technology for common mutations and uncommon mutations that, 14 diseases were not detected by traditional methods. A total of 2,309 (4.7%) carriers of pathogenic mutations in these 14 diseases were found, in all the 320 detected mutations, 63.1% of the mutations are uncommon or never reported. It is supposed that 15.9%–22.3% of the carriers will be missed with traditional carrier screening technique, resulting in an increased risk for children to have genetic diseases [10].

Martin J et al. [11] (2015) published the clinical application of high-throughput sequencing technology to screen the 623 recessive genetic diseases, in Igenomix. The results showed that about 84% (2,161/2,570) of the samples carried at least one pathogenic mutation, about 5% (7/138) couples who were going to undergo assisted reproductive surgery carried a disease-causing gene mutation causing the same genetic disease, 2% (6/287) females had pathogenic mutations in X-linked genetic diseases.

Franasiak JM et al. [12] (2016) analyzed 6,643 individuals (3,738 couples) in infertility centers who were screened for extended carriers of 102–117 diseases. They found 8 couples were carriers of the same genetic disease, showing that they were at risk of having children with a genetic disease, and the impact of extended carrier screening on clinical decisions of infertile patients was as high as 0.21% [12].

3   Carriers Screening System

3.1   Guidance for the Screening

Screening for extended genetic diseases carriers has been investigated by many scientists, and many professional associations in the United States have published a number of guidelines and recommendations for screening for carriers of extended genetic diseases before pregnancy/prenatal.

Due to the rapid development of expanded carrier screening, in 2013, the American College of Medical Genetics and Genomics (ACMG) issued a position statement [14] on prenatal/preconception expanded carrier screening, including criteria for disease selection, recommendations for disease phenotypes and etc. According to the statement, before diagnosis, it is necessary to select the diseases clearly and identify the corresponding mutations causing the diseases, so that to assess fertility risks. The statement also briefly provides guidance on genetic counseling before and after the monogenetic diseases testing.

Further, in 2015, ACMG, American Society of Obstetrics and Gynecology (ACOG), Perinatal Quality Foundation (PQF), American Society for Genetic Counseling (NSGC), and American Society of Maternal and Fetal Medicine (SMFM) issued a joint statement guidance [15] for the screening of extended carriers in reproductive medicine. The guideline states that for women of child-bearing age, carrier screening should be performed before pregnancy, gamete donors should also perform carrier screening before other screening tests. The guideline also explains the screening methods and screening disease standards in different periods.

Haque IS et al. [13] (2016) published a research result of 346,790 individual samples for extended carriers screening of 94 diseases. Among different races/ethnicities, compared with the guidelines recommended by professional associations, Haque IS et al.'s extended carrier screening solution detected more mutations and potential risk fetuses, and the detection rate increased by 9%–55% in different races. In addition, expanding the scope of screening could reduce the difference in screening sensitivity between races, reducing the difference in positive rate between races from 42.2 times to 4.2 times [13].

Following the guidelines, expanded carrier screening can be applied to test for a broad array of genetic conditions before conceiving, regardless of the parents' ethnic or geographical profile. It can screen hundreds of monogenic diseases, so that to know whether the parents are at risk of passing along any of these genetic conditions to their baby.

Our monogenic disorders interpretation system also follows the guidelines for the screening of extended carriers.

3.2 Screening Process

For carrier screening, our monogenic disorders screening system provides detection of 160 recessive genetic diseases which have high morbidity and can severely affect the quality of life and longevity. Through early screening and intervention, we can avoid the happening of the diseases, such as hereditary deafness, albinism, hemophilia, fanconi anemia, phenylketonuria, hepatolenticular degeneration, color blindness and the other diseases.

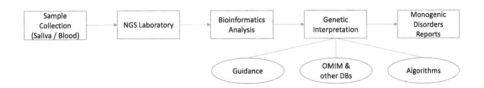

Fig. 1. The service process of monogenic disorders carrier screening.

Figure 1 demonstrates the monogenic disorders carrier screening service process, the sample could be collected through saliva or blood, and saliva could be collected from customer's home and then be posted to NGS laboratory. In the laboratory, the DNA will be extracted and then sequenced by NGS equipment such as NovaSeq [19], the sequencing data will be stored in storage system in fastq format. After quality control, the DNA sequencing data will be analyzed by bioinformatics pipelines. The analyzed DNA sequencing data will further be interpreted by the genetic interpretation system.

The customer will finally receive a monogenic disorders report.

3.3 System Implementation

To implement the monogenic disorders screening service, we should determine the sample collection method, the next generation sequencing experimental platform, the bioinformatics analysis pipeline and running environment, the interpretation database, and the internet service system.

To make the sample collection become easier and non-invasive for customers, we choose to collect saliva samples from customers. Based on the monogenetic diseases to be screened, we customized a panel for sequencing the samples.

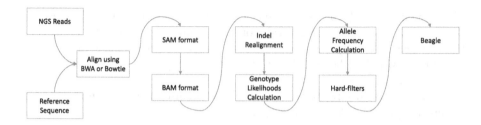

Fig. 2. The indels calling pipeline.

The nucleotide polymorphisms (SNPs) [22] and insertions and deletions (INDELs) [23] calling pipelines were implemented in docker environment, for two reasons: 1) most of the bioinformatics engineers are bothered by the installation and configuration of bioinformatics pipelines, if a pipeline has been installed and configured in docker, then the docker could be reused easily; 2) docker can support the parallel analysis of multiple samples conveniently that each docker could be used to analyze a sample, and we can schedule the running of multiple dockers simultaneously. The indels calling pipeline is described in Fig. 2.

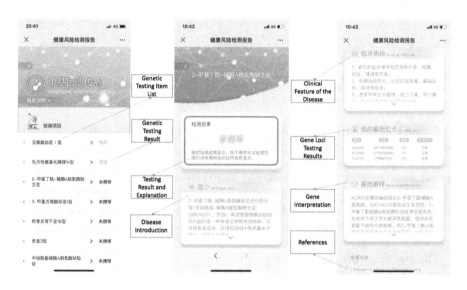

Fig. 3. The homepage and the second-level page of the monogenic disorders testing report.

Another crucial part of the monogenic disorders interpretation system is the genetic interpretation database, which was established combining the OMIM database and the other verified research and clinical results.

Our monogenic disorders screening web service was implemented in Spring cloud [20] environment using spring boot, mybatis, bootstrap, jquery, html5, css, restful api, and mysql. The consumers can access the service through WeChat [21], the genetic interpretation report interfaces are demonstrated in Fig. 3.

Figure 3 left is the homepage of monogenic disorders report, the homepage lists all the 160 monogenic disorders testing items and their corresponding testing results, if the testing result is positive, it will be marked as carrier, otherwise marked as not carried. Click each item, the second-level page will be opened as described in Fig. 3 middle and right, which gives more detail explanation about the testing result, introduces the monogenic disorder, describes the clinical features, shows the related genes and the loci testing results, and finally interprets the genes which can cause the monogenic disorder. The second-level page also lists the references related to the corresponding monogenic disorder.

4   Conclusions

In this paper, we introduce the importance of monogenic disorders screening, we survey the development of monogenic disorders screening technologies and guidance for screening, then we give the implementation of our monogenic disorders screening system, which can detect more than 160 different single gene genetic diseases. Our system can provide convenient service and friendly user interface for customers for screening the monogenic disorders. The monogenic disorders screening service has already been delivered online providing services for tens of thousands of people.

We believe that with the development of genetic testing technologies, the reduction of genetic testing cost, and the increase of public's awareness of pre-pregnancy carrier screening, the transmission of genetic disease-causing genes will be fundamentally blocked, most of the occurrences of single-gene genetic diseases in families will be prevented, and the rare diseases will be even rarer.

Acknowledgment. This work was partially supported by the Science Foundation of Beijing Language and Culture University (supported by "the Fundamental Research Funds for the Central Universities") (20YJ040007, 19YJ040010, 17YJ0302).

References

1. Babar, U.: Monogenic Disorders: An Overview. Int. J. Adv. Res. **5**(2), 1398–1424 (2017)
2. Chen, N., Van Hout, C.V., Gottipati, S., Clark, A.G.: Using Mendelian inheritance to improve high throughput SNP discovery. Genetics **198**(3), 847–857 (2014)
3. Fountain, E.D., Pauli, J.N., Reid, B.N., Palsboll, P.J., Peery, M.Z.: Finding the right coverage: the impact of coverage and sequence quality on single nucleotide polymorphism genotyping error rates. Mol. Ecol. Resour. **16**(4), 966–978 (2016)
4. Christianson, A., Howson, C.P., Modell, B.: March of Dimes: global report on birth defects, the hidden toll of dying and disabled children. March of Dimes Birth Defects Foundation, White Plains, USA (2006)
5. Kaback, M.M.: Screening and prevention in Tay-Sachs disease: origins, update, and impact. Adv. Genet. **44**, 253–65 (2001)

6. Lalonde, E., Albrecht, S., et al.: Unexpected allelic heterogeneity and spectrum of mutations in Fowler syndrome revealed by next-generation exome sequencing. Hum. Mutat. **31**(8), 918–923 (2010)

7. Valencia, C.A., Husami, A., et al.: Clinical impact and cost-effectiveness of whole exome sequencing as a diagnostic tool: a pediatric center's experience. Front. Pediatr. **3**, 67 (2015)

8. Bell, C.J., Dinwiddie, D.L., et al.: Carrier testing for severe childhood recessive diseases by next-generation sequencing. Sci. Trans. Med. **3**(65), 65ra4 (2011)

9. Lazarin, G.A., Haque, I.S., et al.: An empirical estimate of carrier frequencies for 400+ causal Mendelian variants: results from an ethnically diverse clinical sample of 23,453 individuals. Genet. Med. **15**(3), 178–86 (2013)

10. Azimi, M., Schmaus, K., et al.: Carrier screening of 48,761 patients in the IVF setting utilizing next generation DNA sequencing detects common, rare and otherwise undetectable pathogenic variants in prevalent, society-recommended diseases. Fertil. Steril. **103**(2), e17 (2015). Supplement

11. Martin, J.A., et al.: Comprehensive carrier genetic test using next-generation deoxyribonucleic acid sequencing in infertile couples wishing to conceive through assisted reproductive technology. Fertil. Steril. **104**(5), 1286–1293 (2015)

12. Franasiak, J.M., Olcha, M., et al.: Expanded carrier screening in an infertile population: how often is clinical decision making affected? Genet. Med. **18**(11), 1097–1101 (2016)

13. Haque, I.S., Lazarin, G.A., et al.: Modeled fetal risk of genetic diseases identified by expanded carrier screening. JAMA **316**(7), 734–742 (2016)

14. Grody, W.W., Thompson, B.H., et al.: ACMG position statement on prenatal/preconception expanded carrier screening. Genet. Med. **15**, 482–483 (2013)

15. Edwards, J.G., Feldman, G., et al.: Expanded carrier screening in reproductive medicine-points to consider: a joint statement of the American College of Medical Genetics and Genomics, American College of Obstetricians and Gynecologists, National Society of Genetic Counselors, Perinatal Quality Foundation, and Society for Maternal-Fetal Medicine. Obstet. Gynecol. **125**, 653–62 (2015)

16. Ministry of Health: Report on the Prevention of Birth Defects in China (2012). http://www.gov.cn/gzdt/att/att/site1/20120912/1c6f6506c7f811bacf9301.pdf. Accessed 16 May 2020

17. Hamosh, A., Scott, A.F., et al.: Online Mendelian Inheritance in Man (OMIM), a knowledgebase of human genes and genetic disorders. Nucleic Acids Res. **33**(Database issue), D514–7 (2005)

18. OMIM - Online Mendelian Inheritance in Man, an Online Catalog of Human Genes and Genetic Disorders. https://www.omim.org/. Accessed 16 May 2020

19. NovaSeq. https://www.illumina.com/systems/sequencing-platforms/novaseq.html. Accessed 16 May 2020

20. Spring Cloud. https://spring.io/projects/spring-cloud. Accessed 22 May 2020

21. WeChat. https://www.wechat.com/en/. Accessed 22 May 2020

22. Nielsen, R., Paul, J.S., Albrechtsen, A., Song, Y.S.: Genotype and SNP calling from next-generation sequencing data. Nat. Rev. Genet. **12**(6), 443–451 (2011)

23. Mullaney, J.M., Mills, R.E., Pittard, W.S., Devine, S.E.: Small insertions and deletions (INDELs) in human genomes. Hum. Mol. Genet. **19**(R2), R131–6 (2010)

Medical Diagnosis with Machine Learning

A Prediction Model of Gestational Diabetes Mellitus Based on First Pregnancy Test Index

Jianzhuo Yan[1,2,3], Yanan Geng[1,2,3] (iD), Hongxia Xu[1,2,3(✉)], Shaofeng Tan[3,4], Dongdong He[3,4], Yongchuan Yu[1,2], Sinuo Deng[1,2,3] (iD), and Xiaoxue Du[1,2,3]

[1] Faculty of Information Technology, Beijing University of Technology, Beijing 100124, China
xhxccl@bjut.edu.cn
[2] Engineering Research Center of Digital Community, Beijing University of Technology, Beijing 100124, China
[3] Join Lab of Digital Health, Beijing University of Technology and Beijing Pinggu Hospital, Beijing 100124, China
[4] Information Center of Beijing Pinggu Hospital, Beijing 101200, China

Abstract. The purpose of this study is to discuss the possibility of predicting gestational diabetes mellitus (GDM) by analyzing the first test indexes. In order to verify the prediction effect, we used 61 indexes, including age and 60 test indexes, from December 2015 to May 2018 in Beijing Pinggu District Hospital, and conducted experiments of GDM risk prediction based on a variety of different models, ranged from LR, LDA, RF to XGBoost. The experimental results reveal that compared to the dataset of using major relevant indicators, the dataset of using full indicators performs better. Besides, logistic regression can achieve a relatively good prediction effect. On the test set of all data, the area under the curve (AUC) of the Logistic regression model reaches 0.7787. In the meantime, the accuracy rate of the Logistic Regression model reaches $(69.991 \pm 2.833)\%$, and the recall rate and the mean value of the F1 value are $(70.598 \pm 2.210)\%$ and $(70.264 \pm 2.128)\%$, respectively. So the analysis based on the first pregnancy test can play a role in predicting GDM to a certain extent.

Keywords: Predict gestational diabetes mellitus · First test indexes · Logistic regression

1 Introduction

With the improvement of living standards and the aging of China's population, the incidence of diabetes has increased year by year, and it has changed from a rare disease to an epidemic [1]. In 2017, nearly 425 million people worldwide had diabetes [2]. Diabetes is dangerous for pregnant women. When pregnant women give birth under the condition of GDM, the risk of adverse consequences increases for maternal and offspring [3, 4]. For such high-risk GDM, the prevalence rate is also high and rising. In 2017, 21.3 million women had some form of hyperglycemia, of which about 86.4% belonged to GDM [2].

© Springer Nature Switzerland AG 2020
Z. Huang et al. (Eds.): HIS 2020, LNCS 12435, pp. 121–132, 2020.
https://doi.org/10.1007/978-3-030-61951-0_12

In the face of high risk and prevalence of GDM, the traditional medical testing methods mostly use the two-step method at present. First, the sugar screening test is performed on the 24th–28th week after the last menstruation of the pregnant woman. Then if the blood sugar level reaches or exceeds the threshold, doctors will recommend continuing the glucose tolerance test [5, 6]. However, this is relatively late, and long-term hyperglycemia has adverse effects on both offspring and maternal. Early diagnosis and intervention may improve maternal, fetal, and neonatal outcomes [7, 8].

As early as 1959, computer technology was introduced into the medical field, and American medical expert Ledley et al. established the first computer-aided system. Nowadays, machine learning algorithms and classification models are applied in various medical data industries by the advantages of data mining and analysis [9, 10].

In the prediction of diabetes, Alexis Enrique Marcano Cedeño proposes artificial plasticity of multi-layer perceptron as the prediction model of diabetes, and the best result is 89.93% [11]. Calisir and Dogantekin propose a method of Linear Discriminant Analysis Morlet Wavelet Support Vector Machine, which is a system for feature extraction and reduction of diabetes diagnosis using Linear Discriminant Analysis [12]. There are some genetic tests to predict diabetes, but it is too complicated for existing conditions [13]. Under the condition of prior medical knowledge, some people chose a single index or multiple data indexes and used the packaged statistical software such as SPSS software to analyze the test results [14].

Although some articles have proposed methods for predicting GDM, some biomarkers used will not be measured in blood sampling tests. Some articles introduce afamin, and so on as biomarkers [15]. Because these are not necessary items, pregnant women will not test these items because of economic reasons. Someone studies specific indicators. Patrick N A et al. found that mid-upper arm circumference (MUAC) were identified as significant risk factors of GDM [16]. Clive J. Petry et al. found that age at menarche is connected with the future risk of gestational diabetes [17]. So we want to use simple laboratory system indexes so that predictive models can be used in general hospitals.

The logistics regression analysis model is widely used in the area of medicine, and it is also a common method for screening disease indexes [18]. Moreover, a study shows that complex machine learning is no better than logistic regression for clinical prediction models [19]. In this experiment, it is found that the logistic model has a better effect and stability among the four models. For pregnant women with GDM, early test indicators may already be abnormal and develop GDM later [20]. Therefore, we want to construct a predictive model through Logistics regression, use and analyze the hospital's real data set to realize GDM risk prediction. We use the data from the first pregnancy test, which can also be measured in some general hospitals. Pregnant women do not need other tests, which saves money. Finally, according to the predicted results, pregnant women are given dietary intervention in advance, to prevent GDM or reduce maternal and infant injuries. Initially, we need to regularize the hospital's data set and screen the useful laboratory indexes.

The remainder of the paper is organized as follows. We describe the related models in Sect. 2. In Sect. 3, this paper begins to analyze the actual data, constructs predictive models, and compares with other classification methods to verify the validity and stability in different methods and different data set. In Sect. 4, the experimental results are presented. Finally, the paper ends with conclusions and further research topics.

2   Methods

2.1   Logistic Regression Model

Logistic Regression (LR) is the primary model that this paper applied, and it is widely applied in data mining, automatic disease diagnosis, and economic forecasting [21–23]. It is used to solve binomial problems.

In the binomial logistic regression model, the model is conditional probability distribution as follows:

$$F(x) = P(X \le x) = \frac{1}{1 + e^{-(x-\mu)/\gamma}} \tag{1}$$

$$F(x) = P(X \le x) = \frac{1}{1 + e^{-(x-\mu)/\gamma}} \tag{2}$$

Sometimes for convenience, the weight vector and the input vector are recorded as w and x. Then the LR model like this:

$$P(Y = 0|x) = \frac{1}{1 + e^{w \cdot x}} \tag{3}$$

$$P(Y = 1|x) = \frac{e^{w \cdot x}}{1 + e^{w \cdot x}} \tag{4}$$

The probability of the event represents the ratio between the probability of the event happened and not happened. The log odds of the event of the logit function is

$$\text{logit}(p) = \log \frac{p}{1 - p} \tag{5}$$

For LR, it can be obtained from (3) and (4):

$$log(P(Y = 1|x))/(1 - P(Y = 1|x)) = w \cdot x \tag{6}$$

That is mean, the log odds of the output are $Y = 1$ is the linear function of the input x, and this model is the LR model [24].

2.2   Other Comparative Models

GDM prediction is a two-classification problem, and the relative dimensions of medical data are relatively large. We chose several common classification models such as Linear Discriminant Analysis (LDA), Random Forest (RF), and XGBoost to compare with LR. Their brief descriptions are as followed:

LDA: LDA is a classic linear learning method. It is specially used in the binary classification problem. In a given training sample set, LDA tries to project the sample of a straight line, so that the projection points of the same sample are as close as possible and the different sample points are as far away as possible. Then the projection point position determines the classification of the new sample [25, 26].

RF: RF is mainly used for regression and classification. It is an algorithm that integrates multiple trees through the idea of integrated learning. In RF, for each node of the first decision tree, a subset containing a certain number of attributes is randomly selected from the attribute set of the node. Then an optimal attribute is selected from the subset for partitioning. RF is easy to implement and have low computational overhead. And they can show a powerful performance in many practical tasks. The RF can analyze complex interactions between features, and it has a faster learning speed as the number of input variables increases. So RF is appropriate for huge data sets [27].

XGBoost: XGBoost is a tree boosting extensible machine learning model. It is a new type of Bagging algorithm proposed in recent years and has an integrated algorithm library, which has achieved excellent results in many data mining contests. Its core idea is to add trees and split features to grow a tree. Each time a tree is added, it is actually to learn a new function to fit the residual of the last prediction. When we get more than one tree after training, we can get a corresponding leaf node in each tree according to the characteristics of the sample. Each leaf node will correspond to a score, and the final result only needs to add up the corresponding score of each tree to be the predicted value of the sample [28].

3   Experiment

Firstly, we get the first pregnancy test data from the cooperative hospital and the GDM label from desensitization forms. Then we perform simple data preprocessing. Through the selection of models and the analysis of the optimization results, we also summarize the most suitable model for this research. The overall experimental process is shown in Fig. 1.

a. Firstly, under the guarantee of the hospital information department, we collect 61 indexes from several separate electronic medical record systems.
b. Archiving patient data into data sets that can be used by the model.
c. Then we process the data from the previous step by selecting different models.
d. Optimizing parameters to achieve the best results of model processing in the previous step.
e. Recording multiple parameters and model classification effect in multiple data conditions. And going back to step C to replace the model and repeating the C-D step until the selected model is processed.
f. Sorting all models according to test data. And putting a new set of data into the adjusted models to test their generalization ability, that is to say, to simulate the step of validating the effectiveness with new patients in medical experiments. Because the validation data sets have GDM tags, we can score the practicability of the model by the difference between the system's judgment and the tags.

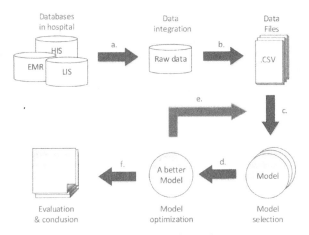

Fig. 1. The overall experimental process

3.1 Data Description

The data are collected from the laboratory examination system of a tertiary hospital in Beijing. The data has been desensitized, and the relevant licenses have been obtained as well. The period spans from December 2015 to May 2018. The data of the first two years are used to construct the training and test data set, and the remaining data is used as the verification set to prevent over-fitting and other situations.

There are 3988 pregnant women in the whole data set. We confirm each pregnant woman's chief complaint and diagnosis through the hospital's medical record system one by one, to ensure the authenticity and validity of the data. Among them, 804 pregnant women have been diagnosed with GDM at least once during pregnancy, while another 3184 were normal pregnant women who have not been diagnosed with GDM throughout pregnancy. Then we select 1,600 subjects (796 normal subjects and 804 diagnosed later). The data situation is shown in Fig. 2.

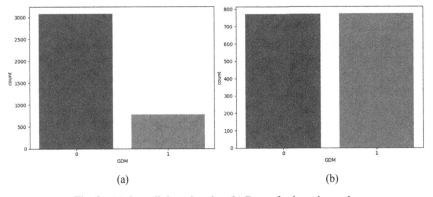

(a)   (b)

Fig. 2. (a) Overall data situation (b) Data of selected samples

We need to digitize the original data and to prepare for the follow-up, as shown in Table 1. In order to make the laboratory indexes more targeted, we add age indexes for each record through the retrieval and verification of the hospital outpatient medical record system. So there are 61 indexes in total. The distribution of age in the overall data and selected samples are shown in Fig. 3.

Table 1. Digital processing and display of category indexes.

Item	Description	Raw data	Digitization
URO	Urinary gallbladder	2+, 1+, +−	3, 2, 1, 0
NaNO2	Nitrite	+, −	1, 0
GLU	Urine glucose	4+, 3+, 2+, 1+, −	4, 3, 2, 1, 0

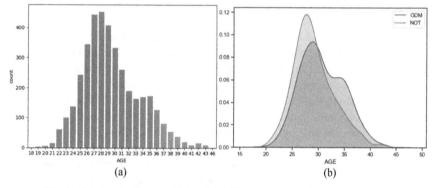

(a)   (b)

Fig. 3. (a) Age in the overall data (b) Different groups of selected samples

For these data, we organize each index's mean of the overall data and the selected 1,600 subjects, as shown in Table 2.

In addition, we adopt the evaluation methods of accuracy, recall, f1-score, and AUC, which are based on the ROC curve.

ROC curve: a receiver operating characteristic curve is a graphical plot that illustrates the diagnostic ability of a binary classifier system as its discrimination threshold is varied [29].

Accuracy: The accuracy measures the overall effectiveness of the classification model, which is the ratio of the positive sample size to the total sample size. The formula is:

$$(\text{Accuracy} = \frac{TP + TN}{TP + TN + FP + FN} \times 100\%) \tag{7}$$

Table 2. Demonstration of data collation of part inspection indexes.

Item	Mean ± Std. Raw data (3988)	Mean ± Std. Patient (804)	Mean ± Std. Normal (796)
GDM	0.202 ± 0.401	1 ± 0	0 ± 0
GLU	4.572 ± 0.508	4.833 ± 0.716	4.511 ± 0.388
CHE	6722.7 ± 1165.962	7163.131 ± 1222.016	6580.079 ± 1119.073
GGT	16.697 ± 11.376	20.689 ± 16.575	15.554 ± 8.356
AGE	29.482 ± 4.094	30.699 ± 4.187	29.146 ± 3.938
WBC	8.774 ± 2.102	9.353 ± 2.214	8.565 ± 1.963
NE#	6.412 ± 1.843	6.872 ± 1.896	6.228 ± 1.723
ALB1	256.582 ± 39.55	267.994 ± 41.306	254.005 ± 39.178
URIC	209.31 ± 51.855	225.193 ± 55.409	205.731 ± 52.435
RBC	4.263 ± 0.351	4.34 ± 0.355	4.23 ± 0.351
ALT	17.481 ± 14.884	21.291 ± 19.316	16.445 ± 13.025
MO#	0.4 ± 0.119	0.425 ± 0.13	0.389 ± 0.109
HCT	38.025 ± 2.627	38.533 ± 2.503	37.838 ± 2.639
PLT	256.084 ± 55.094	268.92 ± 55.52	254.704 ± 53.962
HGB	129.151 ± 10.089	130.879 ± 9.693	128.618 ± 10.062
PCT	0.238 ± 0.047	0.247 ± 0.047	0.237 ± 0.046
FIB	334.235 ± 60.284	344.063 ± 62.614	329.797 ± 59.587
AST	17.319 ± 6.768	18.374 ± 8.874	16.852 ± 5.884

Recall: recall, also known as sensitivity, is the fraction of relevant instances that have been retrieved over the total amount of relevant instances. The formula is:

$$(\text{Recall} = \frac{TP}{TP + FN} \times 100\%) \tag{8}$$

F1-score: the f1-score is a measure of a test's accuracy. It considers both the precision p and the recall r and can fully reflect the performance of the algorithm. The formula is:

$$(\text{F1 score} = \frac{2TP}{2TP + FN + FP} \times 100\% = \frac{2 \cdot \text{Precision} \cdot \text{Recall}}{\text{Precision} + \text{Recall}}) \tag{9}$$

AUC: the area under the curve is equal to the probability that a randomly chosen positive instance is higher than a randomly chosen negative one (assuming 'positive' ranks higher than 'negative') in a classifier [30].

3.2 Model Selection

The strategy we use is first to adopt default parameters, and then optimizes the parameters in the case of grid optimization, so as to achieve the best results. The specific process is shown in Fig. 4.

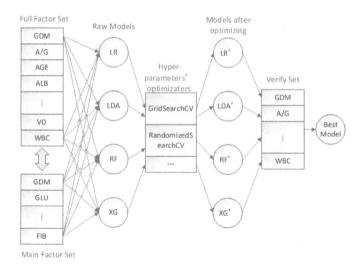

Fig. 4. Model selection process

We use 5-fold cross-validation. The average results are shown in Table 3.

Table 3. Prediction results of 61 inspection indexes prediction by different models.

Model	Accuracy	Recall	F1	AUC
LR	69.991 ± 2.833	70.598 ± 2.210	70.264 ± 2.128	0.7787
LDA	68.707 ± 2.182	67.018 ± 2.861	67.837 ± 2.369	0.7295
RF	65.988 ± 4.155	53.504 ± 2.202	60.549 ± 2.234	0.6997
XGBoost	65.875 ± 1.726	68.644 ± 2.336	67.212 ± 1.705	0.7503

However, in the usual medical literature, some parameters will be screened through prior knowledge, and the main relevant impact indexes will be selected for experiments. Therefore, in order to compare this situation, we also do an experiment using the main related indexes. Then we calculate the correlation between the indexes and the label, then rank them in descending order according to the absolute value of the correlation degree, as shown in Fig. 5.

Relevance calculation: we calculate the correlation between each index and label column separately by the relevance calculation as follow:

$$r(X, Y) = \frac{Cov(X, Y)}{\sqrt{Var[X]Var[Y]}} \tag{10}$$

where the $Cov(X, Y)$ is the covariance of X and Y, Var[X] and Var[Y] are the variances of X and Y.

Selecting the first 15 items as the basis of feature selection, we get the following results as Table 4 through the same process:

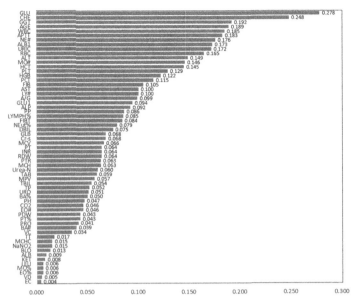

Fig. 5. Model selection process

Table 4. Prediction results of different models under the top 15 high correlation indexes.

Model	Accuracy	Recall	F1	AUC
LR	67.247 ± 4.399	63.710 ± 3.866	65.305 ± 2.934	0.7377
LDA	70.352 ± 2.684	68.204 ± 5.603	69.174 ± 3.681	0.7216
RF	66.097 ± 1.963	57.938 ± 4.141	60.413 ± 2.229	0.6760
XGBoost	66.658 ± 3.013	68.644 ± 2.336	64.800 ± 3.788	0.7382

4 Results

The ROC curves of different data sets and models are shown in Fig. 6.

In this paper, four kinds of models, LR, LDA, RF, and XGBoost, are selected for comparative analysis. From the results, whether using the full indexes set or the main indexes set, the LR model is the most effective and stable. The AUC value of the LR model under the full indexes set is 0.7787. Some medical literature will select the main indexes through prior knowledge, so we also screen 15 indexes with high correlation and compare them with 61 full indexes. By observing the blue and green ROC curves (i.e. 15 indicators and 61 full indicators), it is found that the AUC values of the screening indexes of each model are decreasing. It can be seen that the prediction effect of 61 indicators is much better than the effect of 15 indicators.

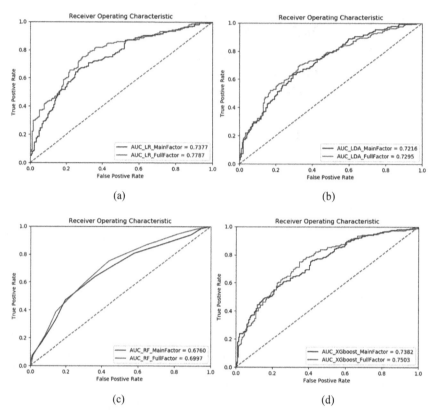

Fig. 6. ROC curves with the main factor and full factor. (a) LR (b) LDA (c) RF (d) XGBoost

5 Discussion

Through the experiments above, we have found that logistic regression analysis has a relatively visible classification effect in this kind of test, and it is easy to operate in optimization and other links. And laboratory indexes can play a predictive effect. Under this method, there are relevant indexes that will bring a more noticeable improvement. In the case of multiple classifiers, multiple tests, and validation sets, the effect of screening using only the main indexes is not as good as in the overall indexes. When we abandon the judgment of prior knowledge based on existing indexes and integrate more indexes into the Logistic model, we may get better results. In this way, while pregnant women undergo the first pregnancy test, they can get a more effective risk assessment of GDM.

Only from the perspective of model construction, this paper discusses the possibility of predicting the GDM incidence using the first pregnancy tests' data. In the future, it can also be used as a plug-in integrated into the endocrine testing system. When a pregnant woman undergoes a preliminary examination, her risk of GDM can be immediately given. And then, doctors can give early risk warning and even dietary intervention. So the prevalence of GDM can be controlled more effectively, helping people save medical costs, maternal time, and money. More importantly, it can reduce the harm to children and pregnant women and ensure their health and safety. In addition, because there is no integrated platform in our hospital, data integration is mostly carried out independently by a separate system. This method applies to this situation. And as a prospect, in the later stage, a more extensive range of indicators can be mined, and multiple case systems can be integrated to improve the prediction accuracy of the model and the effectiveness of preventing GDM.

References

1. Guidelines for the prevention and control of type 2 diabetes in China (2017 ed). Chin. J. Pract. Intern. Med. **38**(04), 292–344 (2018)
2. IDF Diabetes Atlas, 8th ed., International Diabetes Federation, Brussels, Belgium (2017). http://www.diabetesatlas.org/
3. Dornhorst, A., Rossi, M.: Risk and prevention of type 2 diabetes in women with gestational diabetes. Diabetes Care **21**(Suppl.2), B43–B49 (1998)
4. Johns, E.C., Denison, F.C., et al.: Gestational diabetes mellitus: mechanisms, treatment, and complications. Trends Endocrinol. Metabol. **29**(11), 743–754 (2018)
5. American Diabetes Association.: 2. Classification and diagnosis of diabetes. Diab. Care **39**(Suppl.1), S13–S22 (2016)
6. Webber, J., Charlton, M., Johns, N.: Diabetes in pregnancy: management of diabetes and its complications from preconception to the postnatal period (NG3). Br. J. Diab. Vasc. Dis. **15**(3), 107–111 (2015)
7. Thangaratinam, S., et al.: Effects of interventions in pregnancy on maternal weight and obstetric outcomes: meta-analysis of randomised evidence. BMJ **344**, e2088 (2012)
8. Hod, M., Kapur, A., Sacks, D.A., et al.: The International Federation of Gynecology and Obstetrics (FIGO) initiative on gestational diabetes mellitus: a pragmatic guide for diagnosis, management, and care#. Int. J. Gynecol. Obstet. **131**, S173 (2015)
9. Mesa, E., Alberto, J., et al.: Risk of recurrence in operated parasagittal meningiomas: a logistic binary regression model. World Neurosurg. **110**, e112–e118 (2017)
10. Vural, S., Wang, X., Guda, C.: Classification of breast cancer patients using somatic mutation profiles and machine learning approaches. BMC Syst. Biol. **10**, 62 (2016). https://doi.org/10. 1186/s12918-016-0306-z
11. Marcano-Cedeño, A., Torres, J., Andina, D.: A prediction model to diabetes using artificial metaplasticity. In: Ferrández, J.M., Álvarez Sánchez, J.R., de la Paz, F., Toledo, F.Javier (eds.) IWINAC 2011. LNCS, vol. 6687, pp. 418–425. Springer, Heidelberg (2011). https://doi.org/ 10.1007/978-3-642-21326-7_45
12. Caliair, D., Dogantekin, E.: An automatic diabetes diagnosis system based on LDA-Wavelet Support Vector Machine Classifier. Expert Syst. Appl. **38**(7), 8311–8315 (2011)
13. Zhang, H.X., et al.: Research on type 2 diabetes mellitus precise prediction models based on XGBoost algorithm. Chin. J. Lab. Diagn. **22**(03), 408–412 (2018)
14. Cen Liwei, L.W., et al.: The clinical value of fasting plasma glucose, red and white blood cell count combined clinical indicators between 8 and 15 weeks of gestation in the prediction of gestational diabetes mellitus. Prog. Obstet. Gynecol. **28**(3), 182–185+189 (2019)
15. Tramontana, A., et al.: First trimester serum afamin concentrations are associated with the development of pre-eclampsia and gestational diabetes mellitus in pregnant women. Clin. Chim. Acta **476**, 160–166 (2018)
16. Patrick, N.A., et al.: Gestational diabetes mellitus risk score: a practical tool to predict gestational diabetes mellitus risk in Tanzania. Diab. Res. Clin. Pract. **145**, 130–137 (2018)

17. Petry, C.J., Ong, K.K., Dunger, D.B.: Age at menarche and the future risk of gestational diabetes: a systematic review and dose response meta-analysis. Acta Diabetol. **55**(12), 1209–1219 (2018). https://doi.org/10.1007/s00592-018-1214-z

18. Valenzuela, T.D., et al.: Estimating effectiveness of cardiac arrest interventions-a logistic regression survival model. Circulation **96**(10), 3308–3313 (1997)

19. Christodoulou, E., et al.: A systematic review shows no performance benefit of machine learning over logistic regression for clinical prediction models. J. Clin. Epidemiol. **110**, 12–22 (2019)

20. Nanda, S., et al.: Prediction of gestational diabetes mellitus by maternal factors and biomarkers at 11 to 13 weeks. Prenat. Diagn. **31**(2), 135–141 (2011)

21. Li, C.P.: A performance comparison between logistic regression, decision trees and neural network in predicting peripheral neuropathy in type 2 diabetes mellitus. Academy of Military Sciences PLA China (2009)

22. Jiao, X.K., et al.: Logistic regression analysis on hypertension of college students and its correlative factors. Mod. Prev. Med. **39**(19), 3604–3605 (2009)

23. Chi, G.L., et al.: Single factor and logistic multi factor analysis of death during hospitalization in patients with gastrointestinal ulcer bleeding. Med. Innovation China **13**(19), 10–13 (2016)

24. Julien, J., Hoffman, I.E.: Basic Biostatistics for Medical and Biomedical Practitioners, 2nd ed., Tiburon, California (2019)

25. Deng, P., et al.: Linear discriminant analysis guided by unsupervised ensemble learning. Inf. Sci. **480**, 211–221 (2019)

26. Park, C.H., Park, H.: A comparison of generalized LDA algorithms for undersampled problems. Pattern Recogn. **41**(3), 1083–1097 (2008)

27. Wu, H., et al.: Type 2 diabetes mellitus prediction model based on data mining. Inf. Med. Unlocked **10**, 100–107 (2017)

28. Chen, T.Q., et al.: XGBoost: a scalable tree boosting system. In: The ACM SIGKDD International Conference on Knowledge Discovery and Data Mining, pp. 785–794, San Francisco, California, USA (2016)

29. Detector performance analysis using ROC curves - MATLAB & Simulink Example. http://www.mathworks.com. Accessed 11 Aug 2016

30. Fawcett, T.: An introduction to ROC analysis. Pattern Recogn. Lett. **27**(8), 861–874 (2006)

Antisocial Behaviour Analyses Using Deep Learning

Ravinder Singh[✉], Yanchun Zhang, Hua Wang, Yuan Miao, and Khandakar Ahmed

Victoria University, Melbourne, VIC, Australia
ravinder.singh@vu.edu.au

Abstract. Online antisocial behaviour is a social problem and a public health threat. It is one of the ten personality disorders and entails a permeating pattern of violation of the rights of others, and disregard for safety. It prevails online in the form of aggression, irritability, lack of remorse, impulsivity, and unlawful behaviour. The paper introduces a deep learning-based approach to automatically detect and classify antisocial behaviour (ASB) from online platforms and to generate insights into its various widespread forms. Once detected, appropriate measures can be taken to eradicate such behaviour online and to encourage participation. The data for this paper was collected over a period of four months from the popular online social media platform Twitter by using pre-defined phrases linked to antisocial behaviour. Widely used machine learning classifiers: SVM, Decision tree, Random Forest, Linear regression, and deep learning architecture (CNN) were experimented with. CNN was implemented with both GloVe and Word2Vec embeddings and outperformed all the traditional machine algorithms used in the study. Standard performance metrics such as accuracy, recall, precision, and f-measures were used to evaluate classifiers and the CNN-GloVe combination (with 300 dimensions) produced the highest classification performance achieving 98.42% accuracy. Visually enhanced interpretation of the results is presented to demonstrate the inner workings of the classification process.

Keywords: Antisocial behaviour · Deep learning · Machine learning · Behaviour classification · Personality disorder · Twitter mining

1 Introduction

According to the 'Diagnostic and statistical manual of mental disorder' (DSM-5), a diagnostic tool used by mental health professionals [1], there are ten known personality disorders. These disorders are categorized into three clusters based on their similarities and prognosis. Antisocial behaviour is one of the ten personality disorders and falls in a cluster alongside Borderline personality disorder, Histrionic personality disorder, and Narcissistic personality disorder. It is a lasting pattern of behaviour that diverges significantly from the expectations of society. It is usually inflexible, pervasive, and leads to impairment and distress. A person displaying antisocial behaviour violate and disregards the rights of others without considering any implications. Exhibition of

© Springer Nature Switzerland AG 2020
Z. Huang et al. (Eds.): HIS 2020, LNCS 12435, pp. 133–145, 2020.
https://doi.org/10.1007/978-3-030-61951-0_13

antisocial behaviour online appears to be a manifestation of everyday sadism. Online platforms can inadvertently encourage the proliferation of such behaviour by affording culprits access to other online users. Without having any measures in place to restrain such behaviour, online platforms leave a vulnerable group of people at risk. Antisocial behaviour can prevent this vulnerable group of people to lawfully go about their life, and defy them the right for social participation. This paper proposes a natural language and deep learning-based approach, and a conceptual framework that can lay the groundwork to impede and exterminate online antisocial behaviour. We have approached this problem as a text classification task, which includes constructing a benchmark dataset, implementing word embeddings, and training and testing models. Deep learning has been successfully implemented in the past to detect cyberbullying and trolling online [2]. It has been utilized to improve model performance for numerous tasks related to the extraction of unstructured pathology reports [3]. Deep learning has also been used to study adverse drug reactions using social media data [4]. Analogues to our study, the data for this study also focussed on tweets. A study by Nguyen et al. implemented CNNs to successfully capture and classify posts related to crises [5] enabling prompt response. Numerous studies have been conducted, using CNNs, to successfully depict different personalities [6], hay fever [7], mobility patterns [8], and other emotion-related traits [9]. Max-pooling and convolutional operations of CNN enables it to capture even the most salient n-gram features, resulting in the high classification performance. In any text classification task, the dependencies between the words are meaningful and can be effectively used for better performance, and that is where CNN outperforms traditional machine learning algorithms [10]. Taking a leaf from these studies, this paper aims to use CNNs to detect and classify different classes of antisocial behaviour from online posts. The study also aims to empirically validate superior classification performance of CNNs over the traditional machine learning techniques.

1.1 Motivation

Exposure to online antisocial behaviour averts a lot of individuals and inhibits their genuine participation on social media platforms. Ramifications of such behaviour online could propel a vulnerable individual to take extreme actions, and in some instances commit suicide. This study aims to explore and propose a conceptual framework that can be utilized to detect and eliminate antisocial behaviour from online platforms making them safe places for everyone to share their thoughts and learn. The approach can also be extended for detection and prevention of other behaviour and personality disorders online [11].

1.2 Our Approach

The first step in the approach is a binary classification as demonstrated in our previously published worked [12]. This includes capturing relative posts from an online platform, using pre-defined phrases related to antisocial behaviour. These posts are then classified as either antisocial or non-antisocial depending on their contexts. The next step in the process is to further classify antisocial posts into subclasses, each representing a different type of behaviour, such as aggression and disregard for safety, implementing multi-category classification. The approach can benefit in building a system to automatically detect and classify posts related to personality and behaviour disorders, and to take appropriate actions such as deleting the post or banning the author from a platform.

1.3 Contribution

The paper is a work in progress and the following are some of the key contributions so far. A) Construction of a medium-scale antisocial behaviour benchmark dataset with multi-category annotation. B) Classification performance evaluation of machine learning algorithms on our benchmark dataset. C) Classification performance evaluation of Deep learning (CNN) model against the selected machine learning algorithms and empirical validation of its superior performance. D) Insight into the deep learning classification process using visually enhanced interpretations of two popular word embedding variants. E) Online antisocial behaviour knowledge discovery and a proposition of a novel approach to detect and curtail it.

2 Methodology

Figure 1 presents an overview of the conceptual framework and the architecture of our proposed approach to extract posts, classification process, and knowledge generation. Each step is further detailed in the following sub-sections.

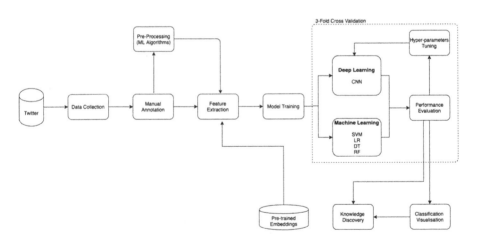

Fig. 1. Conceptual framework for data extraction and classification.

2.1 Data Extraction

The data for this study was extracted from the Twitter platform using NCaputer and Nvivo tools. The posts were captured over a period of four months (November 2019–February

Table 1. Post labels with context.

ID	Post	Context	Label
P1	Now don't waste your time fuelling rumours focus on your fight argo i wish you luck	Non-Antisocial/General	0
P2	I am so fucking sick of seeing christopherhahn he is the biggest most annoying asshole liberal pundit available more than even well maybe not more but close tell him to fuck off and get him off the air	Irritability and aggression	2
P3	I can assure you i do not care for my life attack her and i ll attack something you cherish	Disregard for Safety	3
P4	It's really hard tryna not be an asshole i love pissing people off	Lack of Remorse	4
P5	My niggas in the cage free em out fuck the law i swear when they get out on my life we gonna ball	Failure to conform to lawful behaviour	1

2020) on a regular weekly interval. Pre-defined antisocial behaviour phrases were used to search and capture relative tweets. These pre-defined phrases included aggressive, taboo, unlawful, and insulting words. Not all posts that contained these words were antisocial in nature and some were written in a funny or sarcasm way and these were excluded from the study. A total of 6200 posts were extracted and out of these, about 1500 were omitted due to their confusing context. The posts were captured using Ncapture and were converted into an excel file format using NVivo. Ncaputer runs on Twitter API and hence is bounded by the same rules. The API would only go back a week when searching for posts and would only allow a few attempts to capture these tweets in a day. On some occasions, after conducting several searches with our pre-defined phrases, no relative posts were returned, leaving us with no choice instead to try a week later. A sufficient number of posts were extracted in four months using this web crawling methodology.

2.2 Annotation Process

The complete data set of 4700 twitter posts was annotated by two research students under the supervision of a clinical psychologist. Both these students are active researchers in health informatics and the annotation was performed by using both text of the tweet, and the link to its online repository. The online version was accessed in case there was any confusion related to the context of a tweet and contained any emojis that may have provided further information for the correct interpretation of intent. This methodological approach has been undertaken and has shown superior results in past studies [13]. In case there was no consensus between the annotators related to a post, the post was discarded to avoid misclassification. Cohen's kappa statistic [14] was implemented to calculate the inter-rated reliability of the posts and a significantly high score of k = .84 was achieved. Considering the ethical implications, usernames from the dataset were omitted to keep privacy intact. The posts were categorized into five different classes, with each representing a different aspect and context of antisocial behaviour. Some examples are presented in Table 1. The criteria specified in DSM-5 [1], which is a statistical and diagnostic manual of mental health disorder used by psychologists, was observed. The posts that did not exhibit any signs of antisocial behaviour were labeled as 'General'. The final tally of posts in each class is presented in Table 2.

Table 2. Total number of posts in each class.

Class	Label	Number of posts
Non-AntiSocial/General Tweets	0	1100
Failure to conform to Social Norms concerning lawful behaviour	1	1000
Irritability and aggressiveness	2	1000
Reckless disregard for the safety of self or others	3	800
Lack of remorse, after having hurt someone	4	800

2.3 Model Training and Testing

To achieve the best possible classification results, four different machine learning algorithms were experimented with, namely: Support Vector Machine, Decision Tree, Linear Regression, and Random Forest. The results from these classifiers were then evaluated against the performance of state-of-the-art Deep learning CNNs model. Experiments with traditional machine learning algorithms were conducted with different degree of pre-processing- removing stop words, stemming, both stemming and stop word removal. Stemming only pre-processing produced the superior results with all the four classifiers and these are detailed in the next section. Keeping stop words in the dataset proved beneficial as these often contribute toward context and interpretation of a sentence. For feature extraction, both Term Frequency-Inverse Document Frequency (TF-IDF) and Bag of Words (BoW) techniques were experimented with. Three-fold cross-validation was performed for training and testing classifiers.

In regards to Deep Learning (CNNs), minimal pre-processing was performed on the dataset and mainly entailed omitting, URLs, usernames, emojis, and nonalphanumerical characters. Deep learning algorithms when used with text data, perform their operations on a sequence of words. Retaining stop words and the sequence of words as they appear in a text enables these algorithms to train better by preserving the context depending on the representation of words and their sequence. An approach analogous to sarker et al. [15] has been adopted in this paper. Word-to-vector representation was implemented for feature extraction when implementing CNN. The technique can effectively capture and preserve the subtle relationships among the words in a text enabling superior text classification outcomes. Furthermore, word embeddings unpretentiously extend the size of a feature set, and that is particularly beneficial when working with a small to a medium-sized dataset. The two most commonly used variants of the word embeddings- GloVe and Word2Vec were implemented. The dimension options of 50 (min) and 300 (max) were experimented with. Different hyper-parameters settings were tested. For Optimizer-Nadam RMSProp, Adam, and SGD were tried and for the Activation function- Softmax, relu and sigmoid were tested. Batch sizes of 32, 64, and 256 were used to get the optimal results for the underlying dataset. The model was implemented using Keras, which is an open-source library, and was trained for up to 25 epochs. A comparison study of a maximum number of epochs needed to get the optimal accuracy using the two different word embedding was also performed. Finally, the 3-fold cross-validation, an approach that has been used in numerous previous studies [16, 17], was carried out for training and testing. The standard performance metrics- Accuracy, Recall, Precision, and F-Measure were used for evaluation. Scatter plots and the confusion matrices were generated for both the word embeddings to better understand the classification process. The results of the experiments are discussed in the next section.

3 Results and Discussion

3.1 Accuracy Evaluation

Table 1 shows the classification results for the traditional machine learning algorithms. TF-IDF and BoW, which are two widely used feature extraction techniques, have been implemnted. Three different cases of pre-processing were considered: a) Stemming only; b) Stop word removal only; c) Stemming and stop world removal. This is because machine learning algorithms can produce different results when implemented in different settings. As can be seen from the results, 'Stemming only' pre-processing has resulted in superior performance in all the classifier-feature combinations. This may indicate that stop words play an essential role when machine learning classifiers are used over short text (the average word count of a tweet in our dataset is 20 words). Another observation here is that these classifiers have generated better results when implemented with TF-IDF except for 'DT-BoW' and 'LR- Stemming & Stop Word' combinations. The highest accuracy of 90.82% was achieved by SVM-TF-IDF combination.

The results for the deep learning algorithm CNN, using GoVe and Word2Vec, are presented in Table 3. Different hyper-parameter settings were tested, and the following combination spawned the highest accuracies for both the word embedding. Optimizer: Nadam, Batch_Size: 32, Activation- Function: Relu, Dimension_Setting: 300.

CNN produced higher accuracy when used with GloVe (**98.42**) compared to Word2Vec (94.98). In word2vec, similar words appear co-located together in a vec-tor space and lead arithmetic operations to pose syntactic and semantic relationships. This may work well and generate superior results with large text or certain types of datasets. GloVe, which is a log-bilinear embedding, leads a model to learn relationships explicitly based on co-occurrence matrices, and perform better with a short text and certain type of datasets. Our dataset seems to have fallen into the latter category, leading to GloVe outperforming word2vec (Table 4).

Table 3. Machine learning algorithms performance evaluation.

Model	Feature	Stemming	Stop-word removal	Stemming and Stop-word removal
SVM	TF-IDF	**90.82**	89.42	89.9
DT	TF-IDF	82.3	81.27	81.88
LR	TF-IDF	90.46	88.08	89.34
RF	TF-IDF	84.41	83.85	84.1
SVM	BoW	86.59	84.6	86.02
DT	BoW	82.78	80.61	81.44
LR	BoW	88.22	86.26	89.94
RF	BoW	75.1	73.19	74.01

Table 4. Deep learning performance evaluation.

Model	Feature-Set	Precision	Recall	F-Measure	Accuracy
CNN	GloVe	98.1	98.1	98.1	**98.42**
CNN	Word2Vec	95.2	95.2	95.2	94.98

Confusion matrices in Fig. 2 help to understand the inner workings of both word embeddings on the underlying dataset. In the case of Word2Vec, when used with CNNs, none of the class was identified with 100% accuracy. 'Failure to conform to social norms' (class-1), achieved the highest accuracy of 98%, followed by the general category, which achieved 96% accuracy. The other three classes were identified with a similar accuracy of 93%. Posts falling in classes 2, 3, and 4 were misclassified the most. Three percent of the posts from class-2 were categorized as class-1 posts. Similarly, three percent of class-3 posts were classified as class-4. The biggest confusion among all the classes was 4% of the posts from class-4 were classified as class-2 posts. This might be due to the similarities between words used to express remorse after having hurt someone, to aggressive posts. As can be seen in the word2vec matrix, there is confusion in almost all categories when trying to classify posts. CNN performed better with GloVe and this is demonstrated by the matrix in Fig. 2. 'General' and 'Failure to conform to the social norm' categories almost received perfect scores achieving 99% accuracy. The highest confusion was in class-4 posts that were wrongly and equally classified into all other three classes. Unlike in Word2Vec, the confusion in Glove is spread out into three different classes, instead of concentrated into one main class. None of the classes were misclassified by more than 1%, which is considered as a the good classification performance by any standard.

Scatter plots in Fig. 3 were generated using t-SNE dimensionality reduction technique for visualization. The visualization enables analyzing the classification process from different perspectives. The more defined and separated the clusters are within the

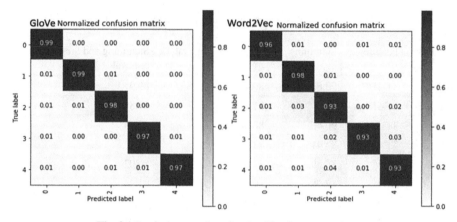

Fig. 2. Confusion matrices for classification comparison.

scatter plots, the more confident was the algorithm in identifying classes. As can be observed from the figure, the clusters are clearly defined and segmented with GloVE embedding indicating superior performance. That does not seem to be the case with word2Vec, which failed to keep a clear distinction among classes. This is indicated by the apparent lack of space among clusters. The overlap can be seen between class-3 and class-4 posts and between class-1 and class 2. There are misclassifications in all classes, however, the embedding was able to works well in identifying the 'General' category as indicated by the relatively well-segmented cluster on the bottom right. Point to note here is that 'aggressive' posts are more similar in nature to the 'failure to conform to social norm' posts and likewise 'disregard for safely' posts possess more similarities to 'lack of remorse' posts, as can be depicted form the scatter plot for word2vec. CNN using GloVe, was able to establish a clear distinction among classes indicating higher probability confidence for each class and the ability to train better on short-text posts.

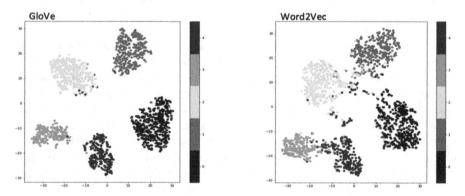

Fig. 3. Scatter plots for classification comparison.

A comparison of training convergence of GloVe and Word2Vec is presented in Fig. 4. The CNN model was trained for 25 epochs with both word embeddings; however, the optimal performance was achieved close to the 15 epoch mark for both the embeddings. It can be noted that GloVe learned faster compared to word2vec indicating superior ability to train on the short text. The number of epochs plays a critical role in the training process. Too little epochs leave a model under-trained, and too many epochs can lead to overfitting. To find the right balance could sometime be a challenging task. The visualization similar to the one in Fig. 4 can assist in establishing the right number of epochs required to train any model. Another thing to consider here is the computing resources and time required for training. The more epochs are required to train a model, the more time and computing resources are used, especially in the case of high dimensionality (e.g. 300 dimensions for this study). So, to find the right number of training epochs is paramount.

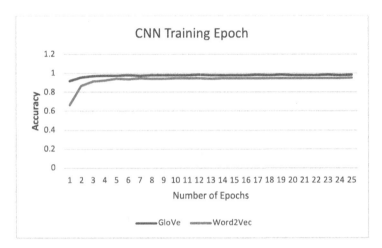

Fig. 4. Training epoch comparison for word embedding variants

The complete results for model evaluation from both machine learning and deep learning classifiers are presented in Table 5. The table shows all the widely accepted evaluation metrics, namely, precision, recall, and f-measures along with performance accuracies. SVM, when used with TF-IDF produced the best result among all the traditional machine learning and feature vector combinations and CNNs, when used with GloVe resulted in the overall highest performance with 98.42% accuracy. Results from other combinations are presented for evaluation purposes.

Table 5. Final algorithm evaluation

Model	Feature-Set	Precision	Recall	F-Measure	Accuracy
CNN	GloVe	98.1	98.1	98.1	**98.42**
CNN	Word2Vec	95.2	95.2	95.2	94.98
SVM	TF-IDF	91.01	91.01	91.01	**90.82**
DT	TF-IDF	82.33	82.33	82.33	82.3
LR	TF-IDF	91.1	90.34	90.68	90.46
RF	TF-IDF	86	84.33	85.17	84.41
SVM	BoW	88.01	86.67	87.34	86.59
DT	BoW	83.67	82.67	83.17	82.78
LR	BoW	89.01	88.34	88.68	88.22
RF	BoW	77.34	74.67	76.01	75.1

3.2 Error Analysis

To understand the classification confusion of algorithms better, some sample misclassified posts are presented in the Table 6. The posts are from CNN + Word2Vec embedding combination. Alongside the actual and predicted labels, the probabilities, which indicate confidence of the algorithm when classifying a post, are also presented. The posts presented are relatively difficult to classify even by human standards due to their confusing contexts. For example, P_1 is a class-4 post, which represents 'lack of remorse', however, it was classified as a class-2 post (aggression). After analyzing the post, it can be inferred that the post possesses both these behaviours, though with a slightly different extent. The confusion of the algorithm is apparent by the low probability score. The words 'I love pissing people off' do represent lack of remorse, however at the same time the words "throw it in your face" represent aggression. Similarly, P_5, which belongs to class-2, was classified as class-0 (general) post. Again, this is a difficult post to classify due to its confusing context and the choice of words, and this is represented by a lower probability score. The words 'thanks to the largest' represent gratitude and points towards the 'General' class, however, the words 'stick it up your ass' points to the contrary and lean more towards 'irritability and aggression'. A few more examples are presented in Table 5. A continuous performance improvement approach can be adopted here by implementing a class-label refinement strategy. Table 7 shows examples of correctly classified posts using CNN-GloVe combination that outperformed CNN-Word2Vec combination.

Table 6. Sample misclassification posts.

ID	Post	Actual label	Predicted label	Probability
P1	I love pissing people off with political correctness lmao if you re going to be an arrogant bigot i m going to throw it in your face	4	2	0.509
P2	Hey anytime fitness stick it up your ass you bunch of crooks this is been cancelled on paper by telephone you re damn right and cancelled by my credit card had to get a new credit card because of you crooks people stay away from anytime fitness scam	2	3	0.527
P3	I will be meeting fady sophie and vivian next month and as with patrick i like to help the fans who can t get to meet them in person so i m doing them cards with comments as well please get your comments for them in quickly as it ll surely be quite a tight squeeze classdw	0	3	0.543
P4	Said this so many times and i ll say it till the day i die i do not care for my life i do not care what happens to me i care about what happens to my friends and i care about my friends lives i want everyone of them to succeed and become great	3	4	0.635
P5	Thanks to the largest tax increase in our nations history you can take that 19th century technology and that tax increase and stick it up your ass	2	0	0.593

Table 7. Sample correctly classified posts

ID	Post	Actual Label	Predicted Label	Probability
P1	Why is it that a passenger in a vehicle can t drink alcohol if they are not driving i ve never understood that like why can t i get fucked up while my homies whippinnn I do anyway	1	1	0.9836875
P3	I reached the top floor at natividad but there was no more helicopters so i jumped from the building and fell into a pool	3	3	0.7692755

4 Conclusion

The state-of-the-art deep learning technique has been implemented and validated against traditional machine learning algorithms to detect and classify antisocial behaviour online. High classification accuracy of 98.42% was achieved using the CNN- GloVe combination outperforming all the other selected machine learning- feature extraction combinations. The rationale behind conducting this study using social media data was that we, now more than ever, spend a tremendous amount of time online engaging with friends and family. Online platforms have taken over most of our daily activities and this is where we exhibit our behaviour and personality traits. In addition to facilitating our day to day activities, these online platforms, unfortunately, have also become breeding grounds for antisocial behaviour in its many disparate forms. We, in this study successfully experimented, and proposed a conceptual framework that can lay the groundwork for automatically detecting and eliminating antisocial behaviour from these platforms. The findings and the results achieved in this study should be considered in light of some limitations. The size of the dataset is relatively moderate in size. This is due to the laborious task of manually annotating each post. Furthermore, not all categories of online antisocial behaviour were considered for this study and the emphasis was laid on the main four categories prevalent online. Lastly, data was extracted only from one online platform-Twitter. For future studies, other classes of antisocial behaviour can be experimented with, by collecting data from more than one online platform. This will also afford the opportunity to work with a larger data set than the one that has been used in this study.

References

1. American Psychiatric Association: Diagnostic and Statistical Manual of Mental Disorder, vol. 5. American Psychiatric Pub, Washington DC (2013)
2. Agrawal, S., Awekar, A.: Deep learning for detecting cyberbullying across multiple social media platforms. European Conference on Information Retrieval, pp. 141–153 (2018)
3. Gao, S.: Hierarchical attention networks for information extraction from cancer pathology reports. J. Am. Med. Inf. Assoc. **25**(3), 321–330 (2018)
4. Nikfarjam, A., Sarker, A., O'Connor, K., Ginn, R., Gonzalez, G.: Pharmacovigilance from social media: mining adverse drug reaction mentions using sequence labeling with word embedding cluster features. J. Am. Med. Inf. Assoc. **22**(3), 671–681 (2015). https://doi.org/10.1093/jamia/ocu041
5. Nguyen, D.T., Mannai, K.A., Joty, S., Sajjad, H., Imran, M., Mitra, P.: Robust classification of crisis-related data on social networks using convolutional neural networks. In; Eleventh International AAAI Conference on Web and Social Media (2017)
6. Majumder, N., Poria, S., Gelbukh, A., Cambria, E.: Deep learning-based document modeling for personality detection from text. IEEE Intell. Syst. **32**(2), 74–79 (2017). https://doi.org/10.1109/mis.2017.23
7. Du, J., Michalska, S., Subramani, S., Wang, H., Zhang, Y.: Neural attention with character embeddings for hay fever detection from Twitter. Health Inf. Sci. Syst. **7**(1), 1–7 (2019). https://doi.org/10.1007/s13755-019-0084-2
8. Singh, R., Zhang, Y., Wang, H.: Exploring human mobility patterns in melbourne using social media data. In: Wang, J., Cong, G., Chen, J., Qi, J. (eds.) ADC 2018. LNCS, vol. 10837, pp. 328–335. Springer, Cham (2018). https://doi.org/10.1007/978-3-319-92013-9_28

9. Poria, S., Chaturvedi, I., Cambria, E., Hussain, A.: Convolutional MKL based multimodal emotion recognition and sentiment analysis. In: IEEE 16th International Conference on Data Mining (ICDM), pp. 439–448 (2016)

10. Gers, F.A., Schmidhuber, J., Cummins, F.: Learning to forget: Continual prediction with LSTM (1999)

11. Islam, M.R., Kabir, M.A., Ahmed, A., Kamal, A.R.M., Wang, H., Ulhaq, A.: Depression detection from social network data using machine learning techniques. Health Inf. Sci. Syst. **6**(1), 1–12 (2018). https://doi.org/10.1007/s13755-018-0046-0

12. Singh, R., et al.: A framework for early detection of antisocial behavior on twitter using natural language processing. In: Barolli, L., Hussain, F.K., Ikeda, M. (eds.) CISIS 2019. AISC, vol. 993, pp. 484–495. Springer, Cham (2020). https://doi.org/10.1007/978-3-030-22354-0_43

13. Colditz, J.B.: Toward real-time infoveillance of Twitter health messages. Am. J. Public Health **108**(8), 1009–1014 (2018)

14. Carletta, J.: Assessing agreement on classification tasks: the kappa statistic. Assessing agreement on classification tasks: the kappa statistic (1996)

15. Sarker, A., Gonzalez, G.: Portable automatic text classification for adverse drug reaction detection via multi-corpus training. J. Biomed. Inf. **53**, 196–207 (2015). https://doi.org/10. 1016/j.jbi.2014.11.002

16. Pandey, D.: Automatic and fast segmentation of breast region-of-interest (ROI) and density in MRIs. Heliyon **4**(12), 1042 (2018)

17. Shin, H.C.: Deep convolutional neural networks for computer-aided detection: CNN architectures, dataset characteristics and transfer learning. IEEE Trans. Med. Imaging **35**(5), 1285–1298 (2016)

A DNN for Arrhythmia Prediction Based on ECG

Yilin Wang[1], Le Sun[1(✉)], Hua Wang[2], Nikita Shklovskiy-Kordi[3], Jun Xu[1], Yongping Lu[4], and Kouzhen Yuan[4]

[1] Nanjing University of Information Science and Technology, Nanjing 210044, China
sunle2009@gmail.com
[2] Sustainable Industries and Liveable Cities, Victoria University, Footscray, Australia
[3] National Center for Hematology, Moscow, Russia
[4] The Second Affiliated Hospital of Henan University of Traditional Chinese Medicine, Zhengzhou, China

Abstract. Arrhythmia is a common disease in the elderly. If not found in time and without effective treatment, it will lead to serious consequences. Electrocardiogram (ECG) is a tool for recording and displaying ECG signals. In this paper, we construct a deep neural network (DNN) based on fully connected neural network to predict the arrhythmia, which has 16 layers. In the DNN, the output layer has four units, corresponding to the normal (N), left bundle branch block (LBBB), right bundle branch block (RBBB), and ventricular premature contraction (VPC) heartbeat respectively. We call the algorithm **E**CG **A**nomaly **P**rediction **D**NN (EAPD). In order to predict the types of the heartbeats that not happened yet, we classify the type of the ECG segment before a heartbeat, rather than classifying the heartbeat itself. We use two time lengths of a segment: 5.6 s and 11.2 s before a heartbeat. Experiment results show that the prediction using 5.6 s segment has better performance.

Keywords: Arrhythmia · DNN · Prediction · ECG

1 Introduction

At present, many researches have conducted on the Arrhythmia detection based on the ECG [2, 4, 6, 8, 10, 11, 14, 15]. This paper focuses on the Arrhythmia prediction problem. We aim to predict the Arrhythmia before it actually happens. In this paper, we divide the ECG segment into four types: normal(N), left bundle branch block (LBBB), right bundle branch block (RBBB), and ventricular premature contraction (VPC) [16]. Figure 1 shows the N, LBBB, RBBB, and VPC heart rhythms.

Supported by the National Natural Science Foundation of China (Grants No 61702274) and the Natural Science Foundation of Jiangsu Province (Grants No BK20170958), and PAPD.

© Springer Nature Switzerland AG 2020
Z. Huang et al. (Eds.): HIS 2020, LNCS 12435, pp. 146–153, 2020.
https://doi.org/10.1007/978-3-030-61951-0_14

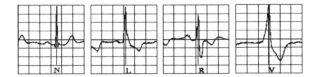

Fig. 1. The heart rhythms of N, LBBB, RBBB, and VPC [16].

We propose a DNN-based model for Arrhythmia prediction, which is called ECG Anomaly Prediction DNN (EAPD). In addition, we do comprehensive experiments to show the efficiency of EAPD on Arrhythmia prediction.

The structure of this paper is: Sect. 2 introduces the related work. Section 3 describes the data preprocessing algorithm and the architecture of the DNN. Section 4 gives the setting of important parameters and shows the results of the experiment. Section 5 summarizes the paper.

2 Related Work

Hannun et al. [3] proposed a DNN that can classify 12 ECG signals, including 10 arrhythmias, one sinus rhythm and one noise. The input data can be raw ECG data without extensive preprocessing. He et al. [7] proposed a pyramid model which can distinguish normal and supraventricular ectopic beats. The model uses the neighborhood information of a heart beat to identify its type. Acharya et al. [1] proposed an 11 layer convolutional neural network (CNN) to classify normal ECG, atrial fibrillation ECG, atrial flutter ECG and ventricular fibrillation ECG.

He et al. [5] proposed a framework of arrhythmia detection with two classification methods named "DHCAF" and "MCHCNN". The former is a feature-based engineering approach and the latter is a deep learning-based approach. Liu et al. [9] proposed a stacked bidirectional LSTM model combined with 2d CNN to extract the overall variation trend and the local characteristics of ECG.

3 ECG Anomaly Prediction DNN

The values of the important parameters are shown in Table. 1. We set up 100 epochs for training and the size of each batch is set to 50. The learning rate and the rate of the Dropout layer are set as 0.001 and 0.3 respectively.

As shown in Fig. 2, the first step is using Pan-Tompkins algorithm [13] to locate QRS wave. This algorithm uses the slope, amplitude and width information to detect QRS waves. The bandpass filter is used to reduce the interference signal. In order to avoid error detection, double threshold detection technique and search back method are used. Then, the samples are intercepted according to the length of the required signals and the location of the QRS. After intercepting a sufficient number of samples, the data sets can be made. Finally, the data sets can be used for training EAPD to obtain the prediction model.

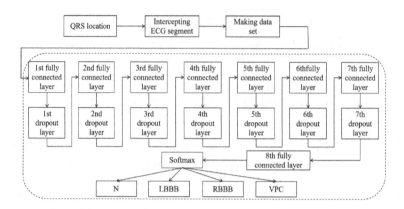

Fig. 2. The architecture of EAPD.

Layer (type)	Output Shape	Param #
dense (Dense)	(None, 2000)	4002000
dropout (Dropout)	(None, 2000)	0
dense_1 (Dense)	(None, 1000)	2001000
dropout_1 (Dropout)	(None, 1000)	0
dense_2 (Dense)	(None, 500)	500500
dropout_2 (Dropout)	(None, 500)	0
dense_3 (Dense)	(None, 250)	125250
dropout_3 (Dropout)	(None, 250)	0
dense_4 (Dense)	(None, 125)	31375
dropout_4 (Dropout)	(None, 125)	0
dense_5 (Dense)	(None, 64)	8064
dropout_5 (Dropout)	(None, 64)	0
dense_6 (Dense)	(None, 32)	2080
dropout_6 (Dropout)	(None, 32)	0
dense_7 (Dense)	(None, 4)	132

Total params: 6,670,401
Trainable params: 6,670,401
Non-trainable params: 0

Fig. 3. Details of EAPD.

The EAPD has 16 layers, including the input and output layers, and 14 hidden layers. The whole prediction process is shown in Fig. 2. The architecture of EAPD is shown in Fig. 3. "Layer (type)" represents the type of each layer of DNN. "Output Shape" represents the shape of the array output from each layer, and "Param #" represents the number of parameters. The parameter here refers to the number of weights of each edge that connects neurons in the current layer to the neurons in the preceding layer. The 15 dense and dropout layers in Fig. 3 correspond to the 15 fully connected and dropout layers in Fig. 2. For example, dense_1 denotes the 2nd fully connected layer, and dropout_1 denotes the 2nd dropout layer. Considering that there are many layers in the network, the activation functions in the hidden layers are all Rectified Linear Unit (Relu). In the output layer, the activation function is Softmax.

Because of the large number of parameters, we add a Dropout layer in each layer to prevent overfitting. The dropout layer randomly sets the output of some nodes in the previous layer to zero, which prevents from updating the weights of these nodes. Randomly deleting some neurons in hidden layers can effectively prevent the occurrence of overfitting and improve the generalization ability of the model.

Table 1. Important parameter settings.

Parameter	Value
Epoch	100
Batch size	50
Learning rate	0.001
Dropout rate	0.3

4   Experiment

We evaluate the EAPD using MIT-BIH arrhythmia database [12]. The experiment runs on a computer with a GPU of NVIDIA GeForce GTX 950M and 3049 MB memory.

4.1   Data Preparation

The MIT-BIH data set contains 48 well-documented double-leaded ECG records. Each record is about 30 min long [12]. Each record is annotated in detail by two or more experts. We use the lead II signal.

The raw ECG data from the MIT-BIH arrhythmia database cannot be used directly in the experiment, they need to be processed to get the desired sample.

After fixing the position of a QRS, we intercept a specific number of sample points left to the QRS. And the type of each heartbeat is the annotation closest to

Table 2. Dimensions of segments A and segments B.

Heart beat type	Data dimension of data set A	Data dimension of data set B
N	(5000, 2000)	(5000, 4000)
LBBB	(5000, 2000)	(5000, 4000)
RBBB	(5000, 2000)	(5000, 4000)
VPC	(5000, 2000)	(5000, 4000)

it. The ECG segments need to be captured in the experiment are the wavebands in front of the N, LBBB, RBBB, and VPC heart beats. We intercept two lengths of data. One is 2000 sampling points taken between the 2100th sampling point to the left of R peak and the 100th sampling point to the left of R peak. The other is 4000 sampling points taken between the 4100th sampling point to the left of R peak and the 100th sampling point to the left of R peak. The length of the former intercepted ECG segment is about 5.6 s, which is the length of about 8 heart beats. The length of the latter is about 11.2 s, which is the length of about 16 heart beats. They are annotated as data sets A and B respectively.

In order to balance the data and avoid the bias of the experiment caused by the number of the ECG segments, we take 5000 segments of each type for both data sets A and B. In the experiment, we scramble the 20000 segments and select 16,000 of them for training and the rest for testing. Besides, we use the One-Hot encoding to encode the labels. The information of data sets A and B are in Table 2. We can see that 5000 samples of each type are taken from A and B, and the number of sample points intercepted in A and B are 2000 and 4000 respectively.

4.2   Result

We use 80% of the data for training. In the training set, we divide 20% of the data into validation sets to verify the algorithm of each epoch. The remaining 20% ECG data are used as a test set.

In data set A, the training accuracy of the first epoch is only 40.22%. But as the number of epochs increases, the accuracy stabilizes and maintains around 97% after the 22nd epoch. In data set B, the training accuracy of the first epoch is only 40.06%. The accuracy maintains around 97% after the 38th epoch. When validating, the accuracy of each epoch produced by the input of the validation set is basically the same as that of the training set. The prediction accuracies of each epoch based on data sets A and B are shown in Fig. 4 and Fig. 5 respectively.

The overall prediction accuracies of the testing sets of data sets A and B are respectively 95.55% and 94.20%. The confusion matrix of data sets A and B are shown in the Table 3 and Table 4, respectively. It can be seen that most signals are correctly predicted.

The *Precision, Recall* and $F_1 Score$ for each category of heart beats for data sets A and B are shown in Table 5 and Table 6 respectively. The overall $F_1 Score$

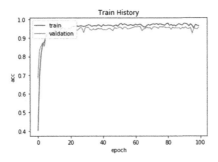

Fig. 4. The training history of data set A.

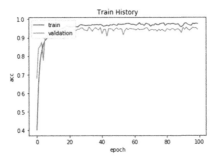

Fig. 5. The training history of data set B.

Table 3. The confusion matrix of data set A.

Confusion matrix		Actual value			
		N	LBBB	RBBB	VPC
Predicting value	N	945	28	2	4
	LBBB	7	987	1	24
	RBBB	7	12	961	19
	VPC	9	51	14	929

Table 4. The confusion matrix of data set B.

Confusion matrix		Actual value			
		N	LBBB	RBBB	VPC
Predicting value	N	952	81	2	4
	LBBB	23	992	0	92
	RBBB	22	2	935	2
	VPC	25	44	8	889

of data sets A and B are 0.96 and 0.93 respectively. The $F_1 Score$ of data set A is 0.03% higher than the F1's of data set B. For data set A, the $F_1 Scores$ of N, LBBB, and VPC are 0.05%, 0.05% and 0.03% higher than those of data set B respectively. The $F_1 Score$ of RBBB of data set A is 0.01% lower than that of data set B.

Table 5. Results of four heart beat types of segments A.

Heart beat type	Precision	Recall	F_1Score
N	96.53%	0.98	0.97
LBBB	96.86%	0.92	0.94
RBBB	96.20%	0.98	0.97
VPC	92.62%	0.95	0.94

Table 6. Results of four heart beat types of segments B.

Heart beat type	Precision	Recall	F_1Score
N	91.63%	0.93	0.92
LBBB	96.88%	0.89	0.89
RBBB	97.29%	0.99	0.98
VPC	91.09%	0.90	0.91

5   Conclusion

In this paper, we propose a DNN called "EAPD" to predict the arrhythmia. Different from diagnosis, prediction can predict the arrhythmia before it occurs. We classify the type of the ECG segment before a heart beat, rather than classifying the heartbeat itself. We use two time lengths of a segment: 5.6 s and 11.2 s before a heartbeat. Experiment results show that the prediction using 5.6 s signal has better performance. In the future, we are going to develop more advanced model to predict more types of arrhythmia and to achieve higher prediction accuracy.

References

1. Acharya, U.R., Fujita, H., Lih, O.S., Hagiwara, Y., Tan, J.H., Adam, M.: Automated detection of arrhythmias using different intervals of tachycardia ECG segments with convolutional neural network. Inf. Sci. **405**, 81–90 (2017)

2. Bajaj, V., Taran, S., Sengur, A.: Emotion classification using flexible analytic wavelet transform for electroencephalogram signals. Health Inf. Sci. Syst. **6**(1), 1–7 (2018). https://doi.org/10.1007/s13755-018-0048-y
3. Hannun, A.Y., et al.: Cardiologist-level arrhythmia detection and classification in ambulatory electrocardiograms using a deep neural network. Nat. Med. **25**(1), 65 (2019)
4. Hassen, H.B., Dghais, W., Hamdi, B.: An e-health system for monitoring elderly health based on internet of things and fog computing. Health Inf. Sci. Syst. **7**(1), 24 (2019)
5. He, J., Rong, J., Sun, L., Wang, H., Zhang, Y., Ma, J.: D-ECG: a dynamic framework for cardiac arrhythmia detection from IoT-based ECGs. In: Hacid, H., Cellary, W., Wang, H., Paik, H.-Y., Zhou, R. (eds.) WISE 2018. LNCS, vol. 11234, pp. 85–99. Springer, Cham (2018). https://doi.org/10.1007/978-3-030-02925-8_6
6. Jagadeeswari, V., Subramaniyaswamy, V., Logesh, R., Vijayakumar, V.: A study on medical internet of things and big data in personalized healthcare system. Health Inf. Sci. Syst. **6**(1), 14 (2018). https://doi.org/10.1007/s13755-018-0049-x
7. He, J., Sun, L., Rong, J., Wang, H., Zhang, Y.: A pyramid-like model for heartbeat classification from ECG recordings. Plos One **13**, e0206593 (2018)
8. Liu, F., Zhou, X., Cao, J., Wang, Z., Wang, H., Zhang, Y.: A LSTM and CNN based assemble neural network framework for arrhythmias classification. In: IEEE International Conference on Acoustics, Speech and Signal Processing (ICASSP), ICASSP 2019, pp. 1303 (2019)
9. Liu, F., et al.: An attention-based hybrid LSTM-CNN model for arrhythmias classification. In: 2019 International Joint Conference on Neural Networks (IJCNN). pp. 1–8 (2019)
10. Liu, F., Zhou, X., Cao, J., Wang, Z., Wang, H., Zhang, Y.: Arrhythmias classification by integrating stacked bidirectional LSTM and two-dimensional CNN. In: Yang, Q., Zhou, Z.-H., Gong, Z., Zhang, M.-L., Huang, S.-J. (eds.) PAKDD 2019. LNCS (LNAI), vol. 11440, pp. 136–149. Springer, Cham (2019). https://doi.org/10.1007/978-3-030-16145-3_11
11. Ma, J., Sun, L., Wang, H., Zhang, Y., Aickelin, U.: Supervised anomaly detection in uncertain pseudoperiodic data streams. ACM Trans. Internet Technology (TOIT) **16**(1), 1–20 (2016)
12. Moody, G.B., Mark, R.G.: The impact of the MIT-BIH arrhythmia database. IEEE Eng. Med. Biol. Mag. **20**(3), 45–50 (2001)
13. Pan, J., Tompkins, W.J.: A real-time QRS detection algorithm. IEEE Trans. Biomed. Eng. **3**, 230–236 (1985)
14. Sun, L.: An extensible framework for ECG anomaly detection in wireless body sensor monitoring systems. Int. J. Sensor Netw. **29**, 101 (2019). https://doi.org/10.1504/IJSNET.2019.10019056
15. Sun, L., He, J., Ma, J., Dong, H., Zhang, Y.: Limited-length suffix-array-based method for variable-length motif discovery in time series. J. Internet Technol. **19**(6), 1841–1851 (2018)
16. Yeh, Y.C., Chiou, C.W., Lin, H.J.: Analyzing ECG for cardiac arrhythmia using cluster analysis. Expert Syst. Appl. **39**(1), 1000–1010 (2012)

Health Behavior and Medication

An Agent-Based Framework for Persuasive Health Behavior Change Intervention

Fawad Taj[1,2](✉) ⓘ, Michel Klein[1] ⓘ, and Aart van Halteren[1] ⓘ

[1] Social AI Group, Vrije Universiteit Amsterdam, Amsterdam, The Netherlands
{f.taj,michel.klein,a.t.van.halteren}@vu.nl
[2] Department of Computer Science, University of Swabi, Swabi, Pakistan

Abstract. With the increased understanding of human behavior and the systematic reporting of a different mechanism through which the behaviors can be influenced, autonomous coaching systems are emerging. In this paper, a generic framework, based on the behavior change ontology, for health behavior change support systems is presented. In the framework, all the intervention components are singly defined and linked. The BDI-based behavior change coaching agent is designed through a framework which delivers behavior change intervention for a variety of coachees (human). Physical activity coaching scenario is created and simulated to show the definition and implementation of different classes in the framework. The simulation results show how through this framework we can define a coaching agent that can apply different types of health behavior interventions across a population of coaches and their preference settings.

Keywords: BDI based agent · Health behavior change · Physical activity · Health behavior framework · Health behavior coach

1 Introduction

The design of digital behaviour change systems and barriers to effective persuasive technologies for a healthy lifestyle are still lacking a unified theory, interdisciplinary awareness, accepted design models, and a common terminology [1]. Recent ontology related to behavior change intervention is defined by combining different classes and their relationships (i.e. Behavior change techniques (BCTs), Mechanisms of Actions (MoAs), behavior, context etc.) gives a common vocabulary and an opportunity to researchers to participate and use ontologies more effectively in their research [2, 3]. In spite of the progress being made, the limited collaboration between technology designers and the health behavior experts usually leads to technologies that are poorly designed or the selection of the health behavior theories are not appropriate or the theory and models selected are not flexible enough to cover all the aspect of the target behavior [4]. The new challenges of this field require new technologies that facilitate the construction of more dynamic, intelligent, flexible and open applications, capable of working in a real-time environment. A generic framework based on the ontologies and theories of behavior change to design a behavior change support systems could be beneficial. The framework

© Springer Nature Switzerland AG 2020
Z. Huang et al. (Eds.): HIS 2020, LNCS 12435, pp. 157–168, 2020.
https://doi.org/10.1007/978-3-030-61951-0_15

would not only provide an opportunity to define design coaching agent but also help design and evaluate a different kind of health behavior interventions.

The recently published ontology for behavior change interventions (BCIO) proposed by S. Michie in *Human behaviour change project* (www.humanbehaviourchange.org) offers a basis for the development of the framework, where the features of an intervention and its relationship were defined. An ontology is a taxonomic description of the concepts in an application domain and the relationships among them. It asserts the characteristics of interventions (i.e., their content and delivery) that are related to behavior through designated mechanisms of action [3].

The framework is generic in the sense that any kind of coaching agent and health behavior change intervention can be defined thought it, because all the common component of behavior change intervention, for example, behavior change techniques (BCTs), mechanism of actions (MoAs), context, mode of delivery etc. are defined and causally linked. All the components within the framework are defined in the section below. Moreover, a BDI-coaching agent published in our earlier paper [5], is implemented as a digital behavior change support agent with a different type of coachee agents, through the framework. The coach identifies strength and weaknesses of the subject, generate a training plan, motivates and helps, just like a human coach. Such a digital coach will highly benefit from the activity tracking system data which is used to personalize the training plan based on performed activities. BDI (Belief, Desire, Intention) agents' paradigm [6] have been presented extensively in the literature, but so far few have elaborated on the adoption of the BDI architecture for health behavior, as proposed in this paper.

Based on this implementation and simulations of digital behavior change support coach and different coachee (human) that it is possible to model a variety of individuals, as well as different types of interventions with multiple active ingredients (i.e. Behavior change techniques). Moreover, the separation of BCTs and MoAs gives an opportunity to plug & play multiple behavioral models from literature which can then be simulated.

The rest of the paper is organized as follows. Section 2 describes the health behavior change support framework and its main components. Section 3 shows the physical activity coach model based on this framework. The implementation of the coach is used in a simulation, and results are described in Sect. 4. Finally, the last section contains a conclusion.

2   Health Behavior Change Support Framework

In this section, we proposed an agent-based generic framework based on the ontology of behavior, which assists the designer to develop and simulate the health behavior change support system with a clear understanding of all the components and the process (see Fig. 1). The main components are the environment & context, intervention, mechanism of action, and the coach and coachee are included to make clear differentiation of each component class. The definition of each of the component is given below.

2.1   Environment and Context

The elements in this component act as input variables and define the intervention and simulation environment for autonomous coaching agent. The context is for defining the

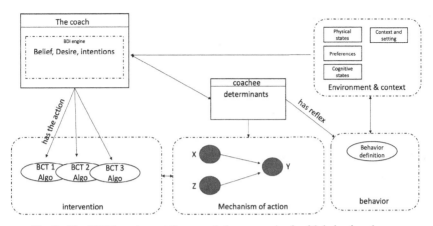

Fig. 1. The BDI-based agent framework for persuasive health behavior change.

target population and setting. Similarly, other variables like coachee (human) preferences physical states etc. can be defined but it all depends on the intervention. For example, in the case of physical activity intervention, motivation, blood pressure, weather, location etc. can be used to define context and environment.

2.2 Intervention

Intervention is a generic term which includes product, service, activity or structural change, intended to achieve behavior change. It can be specified in terms of the content of the intervention and the way this is delivered. In our case, these contents and structure are defined as the function of the coaching agent. For example, in the case of physical activity intervention enforcing self-monitoring, goal-setting, reward delivery etc. can be seen as intervention content.

2.3 Mechanism of Action

Mechanisms of action (MoA) is conceptualized as a range of theoretical constructs, defined broadly as 'the processes through which a behaviour change technique affects behaviour'. These link between BCT and mechanism of action can be easily derived from the heat map given at [7, 8].

Moreover, the links between BCTs and MoAs are important for simulation because the important part in simulation is showing the exact effect of the BCT (i.e., evaluating the processes through which BCTs have their effects).

3 The Physical Activity Coach – Implementation

This section describes how the components defined in the framework above can be used to implement a physical activity coach. BDI architectures have been introduced in several agent-based modelling and simulation (ABMS) platforms. For example, the BDI paradigm integrated into the GAMA modelling platform using its GAML modelling language [9]. We have chosen to implement our coaching agent in GAMA platform.

3.1 Description of the Coaching Scenario

The coaching agent is able to provide three different BCTs as listed in the taxonomy [10] (i.e. 1.1. Goal setting behaviour, 2.2. Feedback on behaviour and, 10.10. Reward on the outcome). According to the framework (see Fig. 1), the effect of these interventions may increase the likelihood of a coachee making an increase physical activity (behavior) by raising motivation and intention about the health risks from not being physically active (mechanism of action). This implementation is composed of 5 steps. The five steps (i.e. coachee and its determinants, environmental & contextual variables, defining intervention, defining mechanism of action, and definition of the BDI coach) are explained with its purpose and explicit formulation.

Creation of Coachee, Environment and Context Variables. We designed a generic behavior model (Fig. 2) as a temporal causal model. The constructs used in this model are based on some well-known established psychological theories of behavior change. For example, self-efficacy and outcome expectations from bandura social cognitive theory; defined as confidence in one's own ability and control over the outcome expected to carry out a particular behavior [11]. Similarly, the health belief model and behavior change wheel define motivation as one of the main construct driving behavior [12, 13]. According to the theory of planned behavior [14], intentions plays an important role to perform behaviors and can be predicted by a number of different kind of behavior constructs and, finally, the goal-setting theory [15] explain the human behavior as "introspective observations" of itself and very much driven by the goal it set for him/herself.

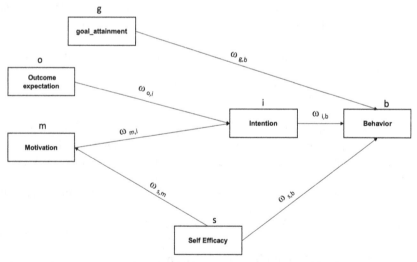

Fig. 2. Generic behavior model for all coachees. The behavior of each coachee will be defined by the construct and connection values.

The motivation and intentions are the compound constructs and they are formalized as follow:

Motivation (m). Motivation is implemented by updating its value with the incoming connection from self-efficacy(s) with a weighted average function to contain it within the range of [0–1]. The symbol $\omega_{s,m}$ represents the connection weight between self-efficacy and motivation. The higher the weight is, the higher would be the impact.

$$m(t + 1) = m(t) + ((1 - m(t)) * s(t) * \omega_{s,m}) \tag{1}$$

Intention (i). The intention is a key driver in coachee behavior and influenced by a number of constructs. The symbol $\omega_{o,i}$ & $\omega_{m,i}$ represents the connection impact of outcome expectation and motivation on intention. It is mathematically formulated as follow:

$$i(t + 1) = i(t) + ((1 - i(t)) * ((e(t) * \omega_{o,i}) * (m(t) * \omega_{m,i})) \tag{2}$$

Behavior (b). The specific target behavior here is daily step count and it depends on a number of socio-ecological constructs for that reason we generate number a random step between 4000 to 6000 and further influenced by the constructs defined in Fig. 2. The formulation is as given below:

$$b(t + 1) = ((i(t) * \omega_{i,b}) + (g(t) * \omega_{g,b}) + (s(t) * \omega_{s,b})) * rand(4000, 6000) \tag{3}$$

We considered three types of agents (human), that are *Red Coachee, Blue Coachee, and Green Coachee*. The initialization for the different coachees and their properties would be defined according to the scenario in the simulation section below. To show the difference, we considered different contextual and intervention variables with different values, which are defined as: window size (9), Percentile (0.6), Total no. days (70), Phase A days (10), and Intervention days (42). The psychological constructs for coachee are intention, out come expectation, self-efficacy, goal attainment, motivation, sensitivity to goal setting, sensitivity to rewards, and sensitivity to feedback on behavior.

Defining Intervention. The simple intervention is applying an adaptive goal setting and reward BCTs to increase daily step counts. The next day goal is based on previous days observed behavior and if the goal is achieved, the coachee is rewarded, otherwise not. To observe the behavior before intervention and calculate average steps count, we established 10 days baseline phase, called as phase A.

Adaptive Goal Setting Algorithm. The adaptive goal-setting algorithm is based on a rank-order percentile algorithm derived from recent developments in basic science around schedules of reinforcement [16]. The algorithm works as follow:

The observed behavior (steps/day) is ranked from lowest to highest and calculation of a new goal based on a *pth* percentile criterion. For example, for one participant, the steps count, each day for their last 9 days (ranked from lowest to highest) was List(L) = 10000, 12000, 13000, 14500, 15000, 15700, 16300,164000, 169000.

The 60th percentile represents a goal of 15700 steps, which becomes the 10th day's goal. Based on [16], the best window size and percentile to consider for physical activity behavior is of 9 and the 60 percentile.

Reward on the Outcome. Rewards are arranged, if and only if there has some effort (or) progress in achieving the behavioral outcomes. The coaching agent delivers a point if and only if the coachee agent achieves the goal set on any certain day.

Defining Mechanism of Actions. The impact of intervention may be affected by the intervention content or mode of delivery and modelling this effect need more research and modelling. For example, some coachee may be reluctant to subject themselves to static goal-setting leading to low physical activity or maybe demotivated with rewards if they found less or unwanted. To make our a very simple mechanism of action we model three different mechanism of action for each of the BCT are explained.

According to [8], the "goal setting" behavior change technique works by instigating the goal attainment and intention process of the person. Similarly, the feedback and reward-based behavior change techniques exhilarate outcome expectation and motivation and self-efficacy and motivation respectively. They are mathematically formulated so that they can be constrained within the range of [0,1].

The behavior model described above predicts the daily steps count of the coachee with the help of psychological constructs. Now when the goal setting is applied, the behavior of coachee changed (means something in the model is changed). That change is presented with these functions.

BDI Based Coaching Agent. The implementation description of the BDI components of the physical activity coach are given below:

Beliefs. Beliefs are represented as predicates in GAML language. Perception is a function executed at each iteration to update the agent's belief base. For example, when the red coachee is perceived, a new belief is added about the steps taken today. The rule is applied when the belief about today step count is updated, a new desire "red_today_steps" is generated.

Goals/Desires. The BDI model assumes that agents are driven by goals. The objectives that the agent would like to accomplish, are updated during the simulation following a goal-plan tree. The desires that the coaching agent can have are represented in Fig. 3.

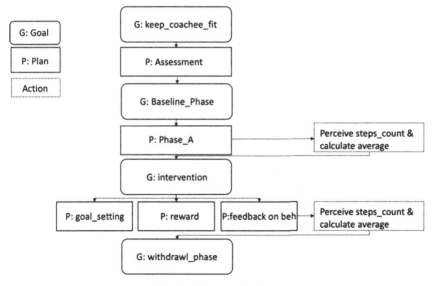

Fig. 3. The goal-plan tree

4 Simulation

Different simulations are performed to investigate the implementation and working of the framework.

4.1 Goal Setting and Reward to Increase Physical Activity

In this part of the simulation, the BDI-based coaching agent applies adaptive goal setting and reward BCT to coachees, to increase their physical activity. All the construct values and connection values are initialized with the same value (e.g. 0.1 and 0.5 respectively), expect the value of 'outcome expectation', which are having values: Red coachee (0.3), Blue coachee (0.5), Green coachee (0.7). As each human are different in nature so as their sensitivity to different intervention. The variation in the sensitivity to goal-setting BCTs is red coachee (0.1), blue coachee (0.4), green coachee (0.9). The values are just assumed randomly for simulation purposes.

The coaching agent applies goal setting and sets an adaptive goal for each next day in the intervention phase and rewarded points to those who achieve its goal. In Fig. 4 it can be seen that the average step count of blue coachee was the highest of three in phase A, but when the goal setting is applied during the intervention phase, the average step count of green coachee increased and became the highest. The increase is due to the high sensitivity of the green coachee to goal setting BCT in comparison to the other two coachees. To see how goal setting actually worked within the coachee (MoA), we can look at the change of the values of their internal concepts. In Fig. 5 it can be seen that goal setting increased the intention and goal attainment determinants of the coachee which leads to increase in their average daily steps count. In Table 1, it can be seen that blue coachee achieved more of its target, so in return, the reward is awarded, which triggered its motivation and self-efficacy (reward mechanism of action). In Fig. 6, it can be seen that the blue coachee self-efficacy and motivation get increased, because he/she achieved more goals. Hence the result shows how different types of coachees, their responses to a different type of intervention, and the effect of these interventions can be simulated.

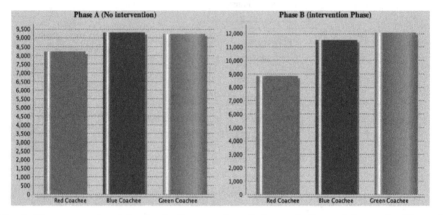

Fig. 4. The average steps taken by the three coachees, before and after applying goal setting and reward BCTs. (Color figure online)

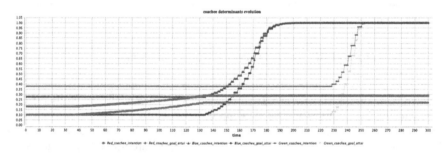

Fig. 5. The evolution of the goal attainment and intention determinants. (Color figure online)

4.2 Feedback on Behavior to Increase Physical Activity

To show how different kind of behavior model can be plugged-in or replaced, we replaced the behavior model, used in Sect. 3.1, with the theory of planned behaviour, which is a well-established social– psychological theory that is used to examine human intentions and behaviour in situations where individuals might lack control over their own behaviour (see Fig. 7). Key construct is behavioral intention, which represents what would motivate and influence users to act in certain behaviours, which in turn influenced by attitude, subjective norm and behavioral control. The best reported BCT for

Table 1. The effect of goal setting on steps count before and after the intervention and the number of times the coachee achieves the goal.

Coachee	Average steps count		Increase percentage	No. of times goal achieved
	Before	After		
Red coachee	8207	8833	7.62%	20
Blue coachee	9315	11503	23.48%	22
Green coachee	9226	12084	30.97%	15

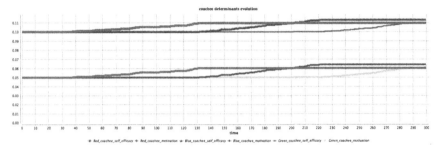

Fig. 6. The evolution of self-efficacy and motivation determinants. (Color figure online)

targeting the "perceived control" determinant is the feedback on behavior. We defined the sensitivities of red, blue, green coachees toward feedback BCT as 0.85, 0.17, and 0.02, respectively.

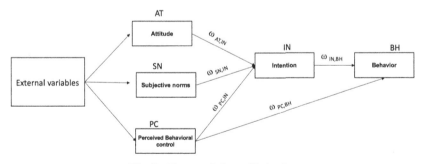

Fig. 7. Theory of planned behavior

In the above coaching agent implementation, we developed our own behavior model that is based on SCT, HBM, and Goal setting theories. This model is basic and is not validated. However, in this framework, several types of models can be used. For example, the COMBI model [17] can be used in the assessment stage to identify the problematic determinants. The identification of problematic determinant(s) can then help in choosing the right kind of behavior change technique in the intervention phase. We used the theory of planned behavior to predict the behavior of the three coachees and with the feedback on their behavior we are trying to influence their control over behavior. Figure 8 shows that the red coachee performed well on feedback, its due to its high sensitivity to feedback and resultantly, Fig. 9 shows the increase in perceived control of the different coachees. So, the result shows how we can plug & play with different kind of theories.

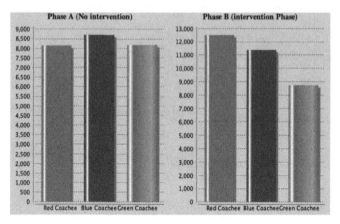

Fig. 8. Average steps count before and after applying feedback BCT. (Color figure online)

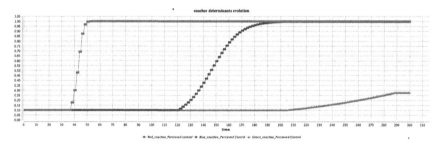

Fig. 9. The evolution of the construct "perceived behavior control" due to intervention. (Color figure online)

5   Conclusion

This work presents the extension and implementation of a BDI based generic coaching agent framework. The framework makes use of the behavior change intervention ontology (BCIO). It has a clear separation between the behavior change techniques and the mechanism of actions through which these techniques work. The working of the framework has been shown by implementing a physical activity coaching agent, which has a strong potential of providing support for different types of individuals with respect to their context and environment. Multiple intervention types of interventions can be defined within the framework, each having their own active ingredients and models.

The coach agent was designed with a specific behavior prediction model and specific BCT to increase daily step counts. Obviously, the model is only a simplified representation of actual human behavior and randomness is used to represent variance between individuals. Human behavior is the interaction between various ecological, psychological, and environmental components, so the framework is so flexible that any type of complex behavior model and separate intervention components and be explicitly defined.

In future, we intend to validate the model by setting up experiments with real users. A mobile or web-based interface would be created to deliver the content of the intervention to the group of people selected for intervention. The experiment would be on the real-time data collected through different mobile sensors or wearables. The intervention design and contents would be developed with the help of the field experts. The intake and the intervention result would be compared with the simulation result for validation. This framework would give us an opportunity to define and validate new hypothesis about the applicability of just-in-time adaptive interventions (JITAI) and also can help vary across types of intervention, contexts and environments.

References

1. Lehto, T.: Designing persuasive health behavior change interventions. In: Wickramasinghe, N., Bali, Rajeev K., Sumoi, R., Kirin, S. (eds.) Critical Issues for the Development of Sustainable E-health Solutions. HDIA, pp. 163–181. Springer, Boston, MA (2012). https://doi.org/10.1007/978-1-4614-1536-7_11

2. Michie, S., West, R.: A Guide to Development and Evaluation of Digital Behaviour Change Interventions in Healthcare. Silverback Publishing, Sutton (2016)

3. Larsen, K.R., et al.: Behavior change interventions: the potential of ontologies for advancing science and practice. J. Behav. Med. **40**(1), 6–22 (2016). https://doi.org/10.1007/s10865-016-9768-0

4. Taj, F., Klein, M.C.A., van Halteren, A.: Digital health behavior change technology: bibliometric and scoping review of two decades of research. JMIR mHealth uHealth **7**(12), e13311 (2019). https://doi.org/10.2196/13311

5. Taj, F., Klein, M., van Halteren, A.: Towards a generic framework for a health behaviour change support agent. In: Proceedings of the 12th International Conference on Agents and Artificial Intelligence, pp. 311–318. SCITEPRESS - Science and Technology Publications (2020). https://doi.org/10.5220/0009173503110318

6. Velleman, J.D., Bratman, M.E.: Intention, plans, and practical reason. Philos. Rev. **100**, 277 (1991). https://doi.org/10.2307/2185304

7. Connell, L.E., et al.: Links between behavior change techniques and mechanisms of action: an expert consensus study. Ann. Behav. Med. **53**, 708–720 (2019). https://doi.org/10.1093/abm/kay082

8. Michie, S., et al.: From theory-inspired to theory-based interventions: a protocol for developing and testing a methodology for linking behaviour change techniques to theoretical mechanisms of action. Ann. Behav. Med. **52**(6), 501–512 (2018). https://doi.org/10.1007/s12160-016-9816-6

9. Taillandier, P., Bourgais, M., Caillou, P., Adam, C., Gaudou, B.: A BDI agent architecture for the GAMA modeling and simulation platform. In: Nardin, L.G., Antunes, L. (eds.) MABS 2016. LNCS (LNAI), vol. 10399, pp. 3–23. Springer, Cham (2017). https://doi.org/10.1007/978-3-319-67477-3_1

10. Michie, S., et al.: The behavior change technique taxonomy (v1) of 93 hierarchically clustered techniques: Building an international consensus for the reporting of behavior change interventions. Ann. Behav. Med. **46**(1), 81–95 (2013)
11. Bandura, A.: Self-efficacy: toward a unifying theory of behavioral change. Psychol. Rev. **84**(2), 191–215 (1977)
12. Janz, N.K., Becker, M.H.: The health belief model: a decade later. Health Educ. Q. **11**(1), 1–47 (1984)
13. Michie, S., Van Stralen, M.M., West, R.: The behaviour change wheel: a new method for characterising and designing behaviour change interventions. Implementation Sci. **6**(1), 42 (2011)
14. Ajzen, I.: The theory of planned behavior. Organ. Behav. Hum. Decis. Process. **50**(2), 179–211 (1991)
15. Locke, E.A., Latham, G.P.: A Theory of Goal Setting & Task Performance. Prentice-Hall Inc., Englewood Cliffs (1990)
16. Adams, M.A.: A pedometer-based intervention to increase physical activity: Applying frequent, adaptive goals and a percentile schedule of reinforcement. UC San Diego (2009)
17. Klein, M., Mogles, N., van Wissen, A.: Why won't you do what's good for you? Using intelligent support for behavior change. In: Salah, A.A., Lepri, B. (eds.) HBU 2011. LNCS, vol. 7065, pp. 104–115. Springer, Heidelberg (2011). https://doi.org/10.1007/978-3-642-25446-8_12

Design Tailored Nutrition and Weight Control Recommendations Using Nutrigenetics and FFQ

Jitao Yang$^{(\boxtimes)}$

School of Information Science, Beijing Language and Culture University,
Beijing 100083, China
yangjitao@blcu.edu.cn

Abstract. The increasing global non-communicable diseases such as obesity and diabetes caused by diet, have roused the urgent needs for simple, modern and tailored solution to achieve effective and healthier lifestyle. Due to the differences in geographical and climatic conditions, different diets appear around the world, and it has long been clear that people don't all respond the same way to the same dietary intervention. One of the ultimate goals of the tailored nutrition is the design of personalized nutritional recommendations to treat or prevent metabolic disorders. Nutrigenetics studies the different phenotypic response to diet depending on the genotype of each individual, and numerous genes and polymorphisms have been already identified as relevant factors in the heterogeneous response to nutrient intake. In this paper, we introduce the knowledge and application of nutrigenetics, and food frequency questionnaire which is used to estimate the frequency of consumption of certain foods, then combining genetic testing data and nutritional intakes, we give the design of tailored nutrition and weight control recommendations. Based on the tailored nutrition recommendations, our system can suggest suitable foods and sports or help customer to place an order to nutrition production factory to produce personalized nutrition supplement products.

Keywords: Nutrigenetics · Tailored nutrition · FFQ · Weight control

1 Introduction

Today's consumers generally have strong sense of health, they are more willing to pay for healthy food and lifestyles. The increasing global diet caused obesity, diabetes and other diseases, have further roused the needs for modern solution to achieve an effective and lasting transition to a healthier lifestyle.

Nutrition is core to human health, however, individual's genes, age, gender, lifestyle, environment and other factors are different, therefore everyone needs different nutrients.

© Springer Nature Switzerland AG 2020
Z. Huang et al. (Eds.): HIS 2020, LNCS 12435, pp. 169–176, 2020.
https://doi.org/10.1007/978-3-030-61951-0_16

With the advancement of nutritional science, nutrigenomics, nutrigenetics, body monitoring equipment and big data analysis technologies, through measuring individual's key physiological indicators, scientists understand the relationship between diet and human metabolism and confirm that there are significant differences in the effects of the same food on individuals. Therefore personalized nutrition becomes necessary to improve our health.

Personalized nutrition has different solutions to leverage human individuality to drive nutrition strategies to prevent, manage, and treat disease and optimize health. In this paper, we propose to use nutrigenomics/nutrigenetics together with food frequency questionnaire (FFQ) to give personalized nutrition and weight control solutions. We first introduce the nutrigenetics and its fast development in recent years, we also describe food frequency questionnaire and its applications, then we implement the personalized nutrition and weight control service. We conclude the paper with the possibility to combine more other biological information to enhance our algorithms for providing personalized nutrition and weight control recommendations.

2   Nutrigenetics

Nutrigenetics has been defined as "the discipline that studies the different phenotypic response to diet depending on the genotype of each individual" [1].

It has long been clear that people don't all respond the same way to the same dietary interventions. For example, some people do really well on a higher fat diet, while others develop high triglycerides or cholesterol on the same diet; some people lose more weight when they reduce carbs, while others lose more weight when they increase complex carbs and reduce fat.

The biological features and functions of organisms are encoded in the genome, which is the heritable material composed by nucleic acids. The genotype is the genetic makeup of an organism, the whole set of genes. Genotype is one of the determinants of the phenotype, the observable characteristics of the organism known as traits. These characteristics range from biochemical and physiological properties to morphology or behavior. The phenotype is also determined by environmental factors together with inherited and non-inherited epigenetic modifications.

Nutrigenetics and nutrigenomics are becoming important areas of scientific and health research globally, which have increased awareness and applications to personalized well-being, as well as public health. After the Human Genome Project [2], there's been a lot of research in nutrigenomics [3].

According to International Society of Nutrigenetics/Nutrigenomics (ISNN) [4], precision nutrition should be considered at three levels: 1) conventional nutritional guidelines into population subgroups by age, gender and other social determinants, 2) individualized nutrition that adds phenotypic information about the person's current nutritional status, 3) genotype-directed nutrition based on rare or common gene variation.

One of the ultimate goals of the promising field of precision nutrition is the design of tailored nutritional recommendations to treat or prevent metabolic disorders [5]. A practical application of nutrigenetics is the use of personal genetic information to guide recommendations for dietary choices that are more efficacious at the individual or genetic subgroup level [6].

3   Food Frequency Questionnaire

The general objective of personalized nutrition is to maintain or improve health by using genetic, dietary and other information to provide more precise and more efficacious personalized healthy eating advice and to motivate appropriate dietary changes.

Personal genetic information could be collected through genetic testing, while dietary information could be collected through food frequency questionnaire (FFQ) [20–22]. FFQ is a common method used to collect dietary data through a list of context-specific food consumption questions, so that to evaluate the relationship between dietary patterns and health outcomes.

There are different commonly used FFQs including: the Harvard University Food Frequency Questionnaire [24], the National Cancer Institute Diet History Questionnaire [25], the Fred Hutchinson Cancer Research Center Food Frequency Questionnaire [26,27], and etc.

Unlike 24-h Dietary Recalls and Weighed Food Records or the other quantitative dietary assessment methods, FFQ usually collects information on the frequency but not the accurate quantity of food consumption, although some FFQs also include usual portion size food list questions. However, FFQ is easier to be implemented and less time-consuming for customers to finish the questions, therefore is relatively easier to be accepted and used by customers.

Generally, FFQ derives the overall nutrient intake by summing all the foods consumed. FFQs have been designed for population groups in the United States, France, China and the other countries in the world.

4   Personalized Nutrition and Weight Control System Implementation

With the identification of polymorphisms or common mutations in vitamin metabolism, large percentages of the population may have higher requirements for specific vitamins [17]. Therefore, to implement personalized nutrition system, genetic testing will be used to detect the genetic features of an individual's response to diet, FFQ will be used to obtain an estimate of nutrition intakes over a time period. Through analyzing genetic testing data and daily food consumption information over a designated time period, our service can provide relative more accurate reports and recommendations for personalized nutrition.

4.1 Genetic Risk Score

Nutrigenetics has evolved from using a unique single nucleotide polymorphism at a candidate gene locus to examine interaction with a specific nutrient (e.g., saturated fat) to a more comprehensive whole genome approach analyzing interactions with dietary patterns [7]. Numerous genes and polymorphisms have been already identified as relevant factors in this heterogeneous response to nutrient intake [8–12]. For example, regarding obesity and metabolic syndrome, recent published studies focusing on gene-environment interactions have revealed important insights that genetic markers have association with metabolic health and fat mass accumulation. These studies also open the door to tailor diets based on individual genetic makeup [13].

Genetic risk score (GRS) [18, 19] was generally derived from the genome-wide association study (GWAS) [28]. For example, the GRS on obesity prediction has been analyzed by [14], and more importantly, the impact of macronutrient intake in the predictive value of this GRS was verified. The GRS in [14] was built as an additive summary measure of a set of 16 genetic variants (according to the number of risk alleles for each variant), including:

- the loci associated with obesity, *i.e.*, rs9939609, FTO; rs17782313, MC4R; rs1801282, PPARG; rs1801133, MTHFR and rs894160, PLIN1; and
- the loci associated with lipid metabolism disturbances, *i.e.*, rs1260326, GCKR; rs662799, APOA5; rs4939833, LIPG; rs1800588l, LIPC, rs328, LPL; rs12740374, CELSR2; rs429358 and rs7412, APOE; rs1799983, NOS3; rs1800777, CETP; rs1800206, PPARA.

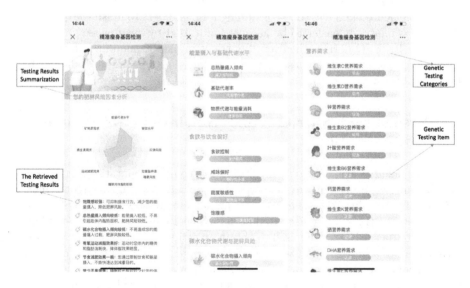

Fig. 1. The homepage of the personalized nutrition and weight control genetic testing report.

The validation of this GRS showed that the high risk group (subjects having more than 7 risk alleles) has increased body mass index ($0.93\,\mathrm{kg/m^2}$ greater BMI), body fat mass (1.69% greater BFM), waist circumference (1.94 cm larger WC) and waist-to-hip ratio (0.01 greater WHR). Therefore, significant interactions between macronutrient intake and GRS prediction values were observed.

Please note that, for different genetic testing items the GRS calculation methods are different.

Our personalized nutrition and weight control service has 8 genetic testing categories: 1) substance metabolism and energy expenditure, 2) appetite and dietary preferences, 3) carbohydrate metabolism and weight control, 4) fat metabolism and weight control, 5) protein metabolism and weight control, 6) the effect of exercise on weight control, 7) the effect of sleep on weight control, and 8) nutrient requirements. Each category includes multiple genetic testing items, such as the nutrient requirement category has more than 10 genetic testing items including Vitamin C requirement, Zinc requirement, and etc.

The homepage of our personalized nutrition and weight control genetic testing report is described in Fig. 1. First, based on the genetic testing results, the report summarizes the customer's genetic characteristics related to weight control and nutrition requirement from 8 dimensions, and for each dimension the report retrieves an important testing item with its testing result and gives brief recommendations. Then, the report lists the 8 genetic testing categories in blocks, each category block has its own testing item list.

Fig. 2. The second-level page of the personalized nutrition and weight control genetic testing report.

Click on the name of each testing item, the second level page will be opened as described in Fig. 2. The first block is the testing result and explanation, which includes the explanation and brief recommendations for the testing result; the second block is the nutrition recommendation which lists the recommended foods for supplementing the required nutrients; the fourth block is the popular science which explains the importance of sufficient supplementation of the tested nutrient; the fifth block illustrates the relationship between the tested nutrient and weight control; the six block is the genes and loci information; the seventh block is the scientific evidence which interprets the genes connected to the tested nutrient; the last block lists the peer reviewed scientific publications.

4.2 FFQ Design

Food frequency questionnaire should be designed for specific population to avoid the missing of important particular foods. Our food frequency questionnaire supports the population groups from different countries, for example, the questionnaire for Chinese is designed based on the dietary structure from the Dietary Guidelines for Chinese [23].

To compute the consumption of different foods' contribution to daily nutrient, the questionnaire has three question categories (*i.e.*, nutrition aims, diet patterns, and nutrition conditions). Different question category includes the corresponding questions such as the diet habit question (*e.g.*, how many portions of fruit were consumed each day), the health condition question (*e.g.*, frequency of catching colds).

Since people generally do not have the patience to finish a long food frequency questionnaire, therefore, we only let people answer less than 40 questions and the questions could be finished in less than 5 min. To make the questionnaire easy to answer and user friendly, all the questions are designed as single choice question or multiple-choice question, there is also a question completion status to let customer know the progress.

Above all, through our personalized nutrition and weight control genetic testing and food frequency questionnaires, we can comprehensively assess the nutritional needs and obesity risks for customers, and provide personalized diet and lifestyle intervention recommendations to help customers to achieve accurate weight control and acquire personalized nutrition.

Additionally, our personalized nutrition and weight control service is connected with nutrition production factory, so that our service can help customer to place an order to let nutrition production factory produce personalized nutrition supplement products.

5 Conclusions

In this paper, through analyzing individual's genetic and food frequency questionnaire data, we implement a personalized nutrition and weight control system. We first explain the importance of nutrition to human health and the necessity of personalized nutrition, since individual's genes, age, lifestyle, environment and other factors are different. Then we introduce the knowledge and development of nutrigenetics and food frequency questionnaire technologies. Finally, we implement the personalized nutrition and weight control system considering and analyzing the genetic testing results and lifestyle data using food frequency questionnaire. Our personalized nutrition and weight control service has been verified and used by many customers, and the feedback indicated that our service can help customers to control weight efficiently and maintain healthy nutrition conditions continuously.

More and more new technologies have enabled multiple endogenous and exogenous factors to be studied at the same time and used to predict the response to nutrition and lifestyle interventions. The endogenous and exogenous factors include epigenomics, metabolomics, microbiomics, and the individual's environment [15] (which is also known as the exposome [16]).

It's planned in our agenda that, more and more endogenous and exogenous data will be included in our algorithms to continuous optimize our personalized nutrition and weight control system.

Acknowledgment. This work was partially supported by the Science Foundation of Beijing Language and Culture University (supported by "the Fundamental Research Funds for the Central Universities") (20YJ040007, 19YJ040010, 17YJ0302)

References

1. Corella, D., Ordovas, J.M.: Nutrigenomics in cardiovascular medicine. Circ. Cardiovasc. Genet. **2**, 637–51 (2009)
2. The Human Genome Project. https://www.genome.gov/human-genome-project. Accessed 9 June 2020
3. McMahon, G., Taylor, A.E., et al.: Phenotype refinement strengthens the association of AHR and CYP1A1 genotype with caffeine consumption. PLoS One **9**, e103448 (2014)
4. Ferguson, L.R., De Caterina, R., et al.: Guide and position of the international society of nutrigenetics/nutrigenomicson personalised nutrition: part 1 - fields of precision nutrition. J. Nutrigenet. Nutrigenomics **9**, 12–27 (2016)
5. Betts, J.A., Gonzalez, J.T.: Personalised nutrition: what makes you so special? Nutr. Bull. **41**, 353–359 (2016)
6. Grimaldi, K.A., van Ommen, B., et al.: Proposed guidelines to evaluate scientific validity and evidence for genotype-based dietary advice. Genes Nutr. **12**, 35 (2017)
7. Frazier-Wood, A.C.: Dietary patterns, genes, and health: challenges and obstacles to be overcome. Curr. Nutr. Rep. **4**, 82–7 (2015)
8. Vallee Marcotte, B.V., Cormier, H., et al.: Novel genetic loci associated with the plasma triglyceride response to an omega-3 fatty acid supplementation. J. Nutrigenet. Nutrigenomics **9**, 1–11 (2016)
9. Ouellette, C., Rudkowska, I., et al.: Gene-diet interactions with polymorphisms of the MGLL gene on plasma low-density lipoprotein cholesterol and size following an omega-3 polyunsaturated fatty acid supplementation: a clinical trial. Lipids Health Dis. **13**, 86 (2014)

10. Rudkowska, I., Perusse, L., et al.: Interaction between common genetic variants and total fat intake on low-density lipoprotein peak particle diameter: a genome-wide association study. J. Nutrigenet. Nutrigenomics **8**, 44–53 (2015)
11. Tremblay, B.L., Cormier, H., et al.: Association between polymorphisms in phospholipase A2 genes and the plasma triglyceride response to an n-3 PUFA supplementation: a clinical trial. Lipids Health Dis. **14**, 12 (2015)
12. Palatini, P., Ceolotto, G., et al.: CYP1A2 genotype modifies the association between coffee intake and the risk of hypertension. J. Hypertens. **27**, 1594–1601 (2009)
13. De Toro-Martin, J., Arsenault, B.J., et al.: Precision nutrition: a review of personalized nutritional approaches for the prevention and management of metabolic syndrome. Nutrients **9**(8), pii:E913 (2017)
14. Goni, L., Cuervo, M., et al.: A genetic risk tool for obesity predisposition assessment and personalized nutrition implementation based on macronutrient intake. Genes Nutr. **10**, 1–10 (2015)
15. Carlsten, C., Brauer, M., et al.: Genes, the environment and personalized medicine: we need to harness both environmental and genetic data to maximize personal and population health. EMBO Rep. **15**, 736–9 (2014)
16. Siroux, V., Agier, L., Slama, R.: The exposome concept: a challenge and a potential driver for environmental health research. Eur. Respir. Rev. **25**, 124–9 (2016)
17. Rozen, R.: Methylenetetrahydrofolate reductase: a link between folate and riboflavin? Am. J. Clin. Nutr. **76**(2), 301–2 (2002)
18. Cooke Bailey, J.N., Igo Jr., R.P.: Genetic risk scores. Curr. Protoc. Hum. Genet. **91**, 1.29.1–1.29.9 (2016)
19. Igo Jr., R.P., Kinzy, T.G., Cooke Bailey, J.N.: Genetic risk scores. Curr. Protoc. Hum. Genet. **104**(1), e95 (2019)
20. Johns, R., Kusuma, J., et al.: Validation of macro- and micro-nutrients including methyl donors in social ethnic diets using food frequency questionnaire and nutrition data system for research (USDA computerized program). SDRP J. Food Sci. Technol. **3**(4), 417–430 (2018)
21. Affret, A., El Fatouhi, D., et al.: Relative validity and reproducibility of a new 44-item diet and food frequency questionnaire among adults: online assessment. J. Med. Internet Res. **20**(7), e227 (2018)
22. Thompson, F.E., Subar, A.F.: Chapter 1 - Dietary Assessment Methodology. In: Nutrition in the Prevention and Treatment of Disease (Fourth edn). Academic Press (2017)
23. Chinese Nutrition Society: Dietary Guidelines for Chinese. People's Medical Publishing House (PMPH). ISBN: 978-7-117-22214-3/R22215 (2016)
24. Rimm, E.B., Giovannucci, E.L., et al.: Reproducibility and validity of an expanded self-administered semiquantitative food frequency questionnaire among male health professionals. Am. J. Epidemiol. **135**(10), 1114–1136 (1992)
25. National Cancer Institute, Division of Cancer Control and Population Sciences: Diet History Questionnaire II (DHQ II) for U.S. & Canada. https://epi.grants.cancer.gov/dhq2/
26. Fred Hutchinson Cancer Research Center: Food Frequency Questionnaires (FFQs). https://sharedresources.fredhutch.org/services/food-frequency-questionnaires-ffq. Accessed 9 June 2020
27. Patterson, R.E., Kristal, A.R., et al.: Measurement characteristics of the women's health initiative food frequency questionnaire. Ann. Epidemiol. **9**, 178–187 (1999)
28. Visscher, P.M., Wray, N.R., et al.: 10 years of GWAS discovery: biology, function, and translation. Am. J. Hum. Genet. **101**(1), 5–22 (2017)

Individualized Medication Guidance Based on Pharmacogenomics

Jitao Yang$^{(\boxtimes)}$ and Bin Li

School of Information Science, Beijing Language and Culture University,
Beijing 100083, China
yangjitao@blcu.edu.cn

Abstract. Adverse drug reactions (ADRs) are a challenge in modern healthcare, particularly due to the increasing complexity of therapeutics, more and more ageing population, and the rising of multimorbidity. Adverse drug reactions are a significant cause of hospitalizations and deaths worldwide. It's estimated that about 1/3 of the patients' death each year worldwide were caused by unreasonable medications. Statistical data show that, in China, about 2.5 million patients were hospitalized each year due to adverse drug reactions, approximately 200,000 people died directly because of unreasonable medication. The incidence of adverse drug reactions in adults is 6.9%, in children this rate is 12.9%, and in newborns this rate is as high as 24.4%. Unreasonable medications happened frequently to newborns, because they are easy to have greater impacts on children, the younger the children are, the more serious the adverse drug reactions are. Therefore, it's very important to find solutions to avoid ADRs occurring to reduce the risk of their harm to patients. Pharmacogenomics is the study of how genes affect a person's response to drugs, therefore, in this paper, based on pharmacogenomics, we established a system to provide individualized medication guidance including dosage recommendations, or withdrawal of the drugs.

Keywords: Adverse drug reactions · Pharmacogenomics · Individualized medication · WDL

1 Introduction

An adverse drug reaction (ADR) can be defined as "an appreciably harmful or unpleasant reaction resulting from an intervention related to the use of a medicinal product; adverse effects usually predict hazard from future administration and warrant prevention, or specific treatment, or alteration of the dosage regimen, or withdrawal of the product" [1].

Adverse drug reactions are challenging the modern healthcare, particularly due to the increasing complexity of therapeutics, the ageing population and the rising of multimorbidity. Adverse drug reactions are a significant cause of hospitalizations and deaths worldwide. It's estimated that about 1/3 of the patients

Z. Huang et al. (Eds.): HIS 2020, LNCS 12435, pp. 177–184, 2020.
https://doi.org/10.1007/978-3-030-61951-0_17

died each year worldwide have unreasonable medications. In China, about 2.5 million patients are hospitalized each year due to adverse drug reactions, approximately 200,000 people died directly because of unreasonable medication. Unreasonable medications were happened frequently to newborns, because they have greater impacts on children, the younger the children are, the more serious the adverse drug reactions are. The incidence of adverse drug reactions in adults is 6.9%, in children, this rate is 12.9%, in newborns, this rate is 24.4%.

Although some ADRs are unpredictable, many are preventable with adequate foresight and monitoring. Preventability or avoidability usually refers to the situation that the drug treatment plan is inconsistent with current evidence-based practice or is unrealistic when taking known circumstances into account [2]. Epidemiological studies tend to find ADRs which are at least potentially preventable. Interventions that reduce the probability of an ADR occurring can be an important way to reduce the risk of patient harm [3].

Generally, we know that diet, gender, and age are important factors in medicine response, in fact genetic factors also affect the respond to medicine. Many drugs currently available are "one size fits all", however they don't work the same way for everyone, and it's difficult to predict who will benefit from a medication, who will not respond at all, and who will experience negative side effects. With the knowledge gained from the Human Genome Project [12], researchers have learned how inherited differences in genes affect the body's response to medications. These genetic differences were used to predict whether a medication will be effective for a particular person and to help prevent adverse drug reactions.

Pharmacogenomics facilitates the identification of biomarkers that can help physicians optimize drug selection, dose, treatment duration, and avert adverse drug reactions. Pharmacogenomics can provide new insights into mechanisms of drug actions and as a result can contribute to the development of new therapeutic agents [4,5].

2 Pharmacogenomics

2.1 Genes and Drugs

Pharmacogenomics is the research on how genes affect a person's response to drugs. This relatively new field combines pharmacology (the science of drugs) and genomics (the study of genes and their functions) to develop effective, safe medications and doses that will be tailored to a person's genetic makeup. Pharmacogenomics uses a patient's genes to foresee how a patient will react to medication, so that to help the patient to avoid unnecessary medications causing side effects and the unnecessary cost on un-effective medications. Health care providers can use pharmacogenomic information to decide the most appropriate treatment for each patient. In addition, pharmacogenomics plays an important role in drug development process, and opens new opportunities in drug discovery.

For instance, Warfarin is a medication used to prevent blood from clotting, genetic variation influences the dose of warfarin to be used for different individuals without causing serious side effects. *CYP2C9* and *VKORC1* are the primary genes affecting patients' different responses to various warfarin doses [24]. Through analyzing the genes, we can help doctor to determine whether a patient's dosing requirement of warfarin may is lower, intermediate, or higher.

Genetic variation can change how much of a medicine is absorbed by our body, how well it gets to where it is needed, and how quickly it gets broken down. If any of these processes are altered, a drug may not work as well for us, or we may have a bad reaction. The chemicals in medications also attach themselves to proteins, affecting potency and how well the medication works once it's in our body system.

Pharmacogenomic testing can provide important information about how genetic variation may affect our response to a medication and suggest ways to change the medicine dose or choice. Pharmacogenomic testing is usually done by taking a small blood or saliva sample. Pharmacogenomic testing can help determine whether a medication may be an effective treatment for patient, what the best dose of a medication is for a patient, whether a patient could have serious side effects from a medication.

2.2 Pharmacogenomics Knowledge Base

Pharmacogenomic testing is not available for all medications, and is available only for certain medications. The Pharmacogenomics Knowledge Base (PharmGKB) [13] is a resource and an interactive tool for researchers to investigate how genetic variation affects drug response [6]. PharmGKB has over 5,000 variant annotations, with over 900 genes related to drugs and over 600 drugs related to genes. The data contained within the PharmGKB database are curated from a variety of sources to bring together the most relevant features of genes, drugs, and diseases for pharmacogenomics [7]. Some data were imported directly from other trusted standard repositories that: 1) gene symbols and names were retrieved from the Human Genome Nomenclature Committee (HGNC) [8,14], 2) drug names and structures were extracted from Drugbank [9,15], 3) detailed relationship data from the literature were manually curated and described using controlled vocabularies.

For genes and drugs where many relationships are known, these are compiled by curators and experts in the field into Very Important Pharmacogene (VIP) summaries (such as [16]) and PharmGKB drug pathways [10], and published in an interactive form on the website and conventional form in peer reviewed journals [11].

2.3 Levels of Evidence

Clinical annotation levels of evidence varies for different drugs and different gene loci, the PharmGKB criteria for levels of evidence is defined in [23].

One single pharmacogenomic testing cannot be used to determine how patient will respond to all medications. Patient may need more than one pharmacogenomic testing if patient is going to take more than one medication.

3   Service Implementation

To implement pharmacogenomic testing service, we selected the drugs that are clinically annotated with clear medication tips and high evidence levels (i.e., Levels 1A, 1B, 2A, 2B [23]) and the corresponding gene loci that are recorded by the databases of genetic pharmacology and pharmacogenomics. A total of 82 drugs in 12 categories were selected to be tested in our system, including common antihypertensive drugs, hypoglycemic drugs, lipid-lowering drugs, gastric ulcer drugs, asthma drugs, gout drugs, antiepileptic drugs, antidepressants, thromboembolic drugs, antipyretics analgesics, antiviral drugs, broad-spectrum antibacterial drugs, etc. According to the guidelines of the clinical pharmacogenomics implementation alliance, based on genetic testing result, a corresponding medication guidance will be given to the customer.

The bioinformatics analysis pipelines were implemented using Workflow Description Language (WDL) [21], which is a user-friendly bioinformatics pipeline scripting language. Most of the bioinformatics pipelines are chained by many software which maybe were developed by Perl, Python, R or the other programming languages. Since the software in each pipeline were developed by different persons or organizations using different languages, the input or output data formats are various, therefore it's very difficult to compile a pipeline successfully. We use WDL to implement our pharmacogenomic pipelines so that to make the software in the pipelines loose coupled and let the pipelines support multiple cluster environment.

WDL has a few very important concepts: workflow, task, call, command and output, runtime, parameter_meta, and meta. Workflow is an executable process, and is composed by multiple tasks, task could be called by the command *call*. Task is composed by *command* block and *output* block, *command* block includes the commands to run in compute nodes, *output* block defines the output of a task. A simple workflow is demonstrated as follows:

```
#define workflow PharmWDL
workflow PharmWDL {
   #define variables
   File pharmFile
   String pharmSample
   #call the task PharmTaskA
   call PharmTaskA{
      input:
      fastaFile=pharmFile
      pharmSampleName=pharmSample
   }
   call PharmTaskB {
```

```
      input:
      inTaskB=PharmTaskA.outRawVCFFile
   }
}
#define task PharmTaskA
task PharmTaskA {
   File gatkFile
   File fastaFile
   File indexFile
   File dictFile
   String pharmSampleName
   File inputBAMFile
   File bamIndexFile
   #define command
   command {
      java -jar ${gatkFile} \
      -T PharmTask \
      -R ${fastaFile} \
      -I ${inputBAMFile} \
      -o ${pharmSampleName}.raw.indels.snps.vcf
   }
   #define runtime
    runtime {
      docker: "pharmacogenomics/Pharm_image"
   }
   #define output
   output {
      File outRawVCFFile = "${pharmSampleName}.raw.indels.snps.vcf"
   }
}
#define task PharmTaskB
task PharmTaskB {
   File inTaskB
   command {
      programB in=${inTaskB} out1=outputFileB1.ext out2=outputFileB2.ext
   }
   output {
      File outFile1 = "outputFileB1.ext"
      File outFile2 = "outputFileB2.ext"
   }
}
```

In the above codes, PharmWDL is the name of workflow, PharmWDL defines two variables pharmFile and pharmSample, and the task PharmTaskA, and PharmTaskB, are called through the command *call* respectively. In the meanwhile, through *input* command, the variables fastaFile and pharmSampleName in the task PharmTaskA are assigned values pharmFile and pharmSample respectively. The task PharmTaskA defines seven variables, the *command* block, and the output of the task. PharmTaskA and PharmTaskB are chained to form a pipeline that, PharmTaskB's input file is the output file of PharmTaskA. WDL supports multiple task combination modes to form different pipelines.

Using WDL and the workflow management system Cromwell [22], our pharmacogenomic pipelines can analyze thousands of samples in parallel conveniently. Some of the tasks could also be reused as other workflows' tasks.

Fig. 1. The homepage and the second-level page of the individualized medication testing report.

The pharmacogenomic interpretation database were created primarily based on PharmGKB (Pharmacogenomics Knowledge Base) [6] and the clinical literature. The interpretation system was developed following the guidelines of CPIC (Clinical Pharmacogenetics Implementation Consortium) [17], DPWG (Dutch Pharmacogenetics Working Group) [19], and Technical Guidelines for the Genetic Testing of Drug Metabolizing Enzymes and Drug Targets [20].

After genetic testing, the customer will get an individualized medication testing report, which includes genetic testing results and medication guidance. The interfaces of the pharmacogenetic testing report are demonstrated in Fig. 1.

Figure 1 left is the homepage of the report, in which the left column is the 12 drug categories, click each drug category, the second column will be shown to list the drug items included in the corresponding drug category, the third column is the genetic testing result with short medication guidance.

Click each drug item, the second level page like Fig. 1 middle and right will be appear: 1) the drug name will be shown in the top of the second level page, 2) below the drug name is the first text block which gives the genetic testing result and the corresponding medication guidance, including dosage recommendations, or withdrawal of the drugs, 3) the second text block introduces the drug functions and dosage information, 4) the third text block gives the contraindication of the drug, 5) the fourth text block lists the testing genes and the loci testing result, 6) the fifth text block explains the relationship between genes and drug, 7) the references are attached at the end of the page.

The individualized medication testing report system was implemented using Java web technologies, and the system was imbedded in WeChat app that, the customer will receive wechat notification when the genetic testing report comes out. We chose to use wechat as a platform to provide the service, because wechat is the most popular messenger app in China that people use it everyday, therefore it is the easiest way for our service to contact and communicate with customers, it's also very convenient for customers to access our service.

Above all, through collecting a customer's saliva, our laboratory can extract customer's DNA from saliva, sequence the DNA, analyze the DNA, interpretate the DNA testing result, and return a professional individualized medication guidance report to the customer.

4   Conclusions

Pharmacogenomics is a new field of medicine that will help doctors to take into account detailed specifications about each patient, resulting in more accurate treatment plans and fewer side effects for patients. In this paper, we introduces the knowledge of adverse drug reaction and pharmacogenomics, the development of pharmacogenomics, and the technologies used for pharmacogenomics; we give the pharmacogenomics knowledge base which are used for interpreting the genetic testing result; we also indicate the clinical annotation levels of evidence which sets the standards for implementing pharmacogenomics services. We describe how to use WDL to implement the pharmacogenomic pipelines to analyze the genetic data. Finally, we demonstrate the individualized medication guidance report for customers, including dosage recommendations, or withdrawal of the drugs. The individualized medication testing service has been delivered online and has been providing services for many customers.

We believe that, with continued research and development, the number of diseases for which pharmacogenomics can develop treatment plans is expected to grow.

Acknowledgment. This work was partially supported by the Science Foundation of Beijing Language and Culture University (supported by "the Fundamental Research Funds for the Central Universities") (20YJ040007, 19YJ040010, 17YJ0302)

References

1. Aronson, J.K., Ferner, R.E.: Clarification of terminology in drug safety. Drug Saf. **28**, 851–70 (2005). https://doi.org/10.2165/00002018-200528100-00003
2. Ferner, R.E., Aronson, J.K.: Preventability of drug-related harms-part I: a systematic review. Drug Saf. **33**, 985–994 (2010). https://doi.org/10.2165/11538270-000000000-00000

3. Coleman, J.J., Pontefract, S.K.: Adverse drug reactions. Clin. Med. **16**, 481–5 (2016)
4. Manolio, T.A.: Genomewide association studies and assessment of the risk of disease. N. Engl. J. Med. **v363**, 166–176 (2010)
5. Wang, L., Howard, L., et al.: Genomics and drug response. N. Engl. J. Med. **364**, 1144–1153 (2011)
6. Thorn, C., Klein, T., Altman, R.B.: PharmGKB: the pharmacogenomics knowledge base. Methods Mol. Biol. **1015**, 311–320 (2013)
7. Altman, R.B., Klein, T.E.: Challenges for biomedical informatics and pharmacogenomics. Ann. Rev. Pharmacol. Toxicol. **42**, 113–133 (2002)
8. Povey, S., Lovering, R., Bruford, E., Wright, M., Lush, M., Wain, H.: The HUGO gene nomenclature committee (HGNC). Hum. Genet. **109**(6), 678–680 (2001). https://doi.org/10.1007/s00439-001-0615-0
9. Wishart, D.S., Knox, C., et al.: DrugBank: a comprehensive resource for in silico drug discovery and exploration. Nucleic Acids Res. **34**, D668–D672 (2006)
10. Eichelbaum, M., Altman, R.B., et al.: New feature: pathways and important genes from PharmGKB. Pharmacogenet. Genomics **19**, 403 (2009)
11. Sangkuhl, K., Berlin, D.S., et al.: PharmGKB: understanding the effects of individual genetic variants. Drug Metab. Rev. **40**, 539–551 (2008)
12. The Human Genome Project. https://www.genome.gov/human-genome-project. Accessed 22 May 2020
13. Pharmacogenomics Knowledgebase. https://www.pharmgkb.org/. Accessed 31 May 2020
14. HUGO Gene Nomenclature Committee. https://www.genenames.org/. Accessed 31 May 2020
15. DrugBank. https://www.drugbank.ca/. Accessed 31 May 2020
16. Lamba, J., Hebert, J.M., et al.: PharmGKB summary: very important pharmacogene information for CYP3A5. Pharmacogenet. Genomics **22**(7), 555–8 (2012)
17. Clinical Pharmacogenetics Implementation Consortium. https://cpicpgx.org/. Accessed 31 May 2020
18. Pharmacogenomics Research Network. https://www.pgrn.org/. Accessed 31 May 2020
19. Dutch Pharmacogenetics Working Group (DPWG). http://upgx.eu/guidelines/. Accessed 31 May 2020
20. Technical Guidelines for the Genetic Testing of Drug Metabolizing Enzymes and Drug Targets. http://www.nhc.gov.cn/yzygj/s7659/201507/03e00d45538d43babe62729a8f635ff7.shtml. Accessed 31 May 2020
21. Workflow Description Language (WDL). https://github.com/openwdl/wdl. Accessed 31 May 2020
22. Cromwell-Workflow Management System. https://github.com/broadinstitute/cromwell. Accessed 6 June 2020
23. Whirl-Carrillo, M., McDonagh, E.M., et al.: Pharmacogenomics knowledge for personalized medicine. Clin. Pharmacol. Ther. **92**(4), 414–7 (2012)
24. Schwarz, U.I., Ritchie, M.D., et al.: Genetic determinants of response to warfarin during initial anticoagulation. N. Engl. J. Med. **358**(10), 999–1008 (2008)

Author Index

Printed in the United States
By Bookmasters